A Multi-Dimensional Approach

Asparaginase

Rajeswara Reddy Erva

A Multi-Dimensional Approach Towards the Novel Bacterial Asparaginase

LAP LAMBERT Academic Publishing

Imprint

Any brand names and product names mentioned in this book are subject to trademark, brand or patent protection and are trademarks or registered trademarks of their respective holders. The use of brand names, product names, common names, trade names, product descriptions etc. even without a particular marking in this work is in no way to be construed to mean that such names may be regarded as unrestricted in respect of trademark and brand protection legislation and could thus be used by anyone.

Cover image: www.ingimage.com

Publisher:
LAP LAMBERT Academic Publishing
is a trademark of
International Book Market Service Ltd., member of OmniScriptum Publishing Group
17 Meldrum Street, Beau Bassin 71504, Mauritius

Printed at: see last page
ISBN: 978-613-9-45733-5

Zugl. / Approved by: Warangal, NIT Warangal, 2016

Copyright © Rajeswara Reddy Erva
Copyright © 2019 International Book Market Service Ltd., member of OmniScriptum Publishing Group

Dedicated to my better half Jayanthi,

my lovely son Ruthweep Reddy

and my beloved Parents

1

Acknowledgements

To begin with most influencing factor of this piece of work, it is my proud privilege to express my deep sense of reverence and gratitude to my supervisor, **Dr. R. Satish Babu** for his excellent supervision, skilled guidance, unfailing support, stimulating discussions, critical evaluation and constant encouragement during this dissertation work. I have learned a lot and improved my skills under his valuable guidance. I thank him for his time and efforts for making me a better researcher. During my Ph.D. tenure he has been more than a guide and a source of inspiration. I am extremely thankful to him for spending his valuable time in spite of his busy schedule.

Part of this research was carried out at Azyme Biosciences, Bangalore, India. I am thankful for all the staff of Azyme Biosciences who helped me to carry-out experimental investigations during enzyme purification.

I also express my sincere thanks to all my fellow scholars and other student friends with whom I laughed my heart out during my stay at NIT Warangal. I am also greatly thankful to all the teaching and non-teaching staff of Biotechnology department, NIT Warangal for their co-operation.

Lastly, I owe a special thanks to my wife, son, parents and well-wishers for their love and encouragement without which I never would have imagined myself to pull it this far.

Rajeswara Reddy Erva

CONTENTS

4

LIST OF FIGURES

LIST OF TABLES

ABBREVATIONS

Symbols	
#	Number
%	Percentage
®	Registered trade-mark
°C	Temperature in degree Centigrade
°K	Temperature in degree Kelvin
R^2	Correlation Coefficient

Weight and Measures	
Da	Daltons
gm	Gram
h	Hour/Hours
IU	International Unit
IU/ml (or) $IUml^{-1}$	International Unit per milliliter
Kcal	Kilocalories
Kcal/mol	Kilocalories per mole
kDa	Kilo Daltons
KJ	Kilojoules
KJ/mol	Kilojoules per mole
$KJ\ mol^{-1}\ nm^{-1}$	Kilojoules per mole per nanometer
L	Liter
Min	Minute
ml	Milliliter
M	Molar/Molarity

MMol	Mole
N	Normal/Normality
nm	Nano Meter
ns	Nano Seconds
rpm	Rotations Per Minute
μg	Microgram
μg/ml	Microgram per milliliter

Concentration

v/v	Volume/Volume
w/v	Weight/Volume

Text Abbreviations

1NNS	PDB ID of L-asparaginase from *Escherichia coli*
A	Absorbance
AC	Ammonium Chloride
ALL	Acute Lymphoblastic Leukemia
ANN	Artificial Neural Networks
ANOVA	Analysis of Variance
APS	Ammonium Persulphate
BLAST	Basic Local Alignment Search Tool
BP	Back-Propagation
BSA	Bovine Serum Albumin
CBB	Coomassie brilliant blue
CCD	Central Composite Design

11

CPU	Central Processing Unit
CV	Coefficient of Variation
DAHP	Di Ammonium Hydrogen Phosphate
DDR	Double Data Rate
DEAE	Diethyl Amino Ethyl
DO	Dissolved Oxygen
DPHP	Di Potassium Hydrogen Phosphate
DNA	Deoxyribonucleic acid
EA*KCTC 2190*	*Enterobacter aerogenes KCTC 2190*
ERRAT	A program for verifying protein structures
FCCCD	Face Centered Central Composite Design
FDA	Food and Drug Administration
FFT	Fast Fourier Transform
FS	Ferrous Sulphate
GA	Genetic Algorithm
GROMACS	GROningen MAchine for Chemical Simulations
GS	Glutaminase Synthase
H-bond	Hydrogen Bond
HL-60	Human promyelocytic leukemia cells
HMM	Hidden Markov Model
Inn Age	Inoculum Age
Inn Size	Inoculum size
KCTC	Korean Collection for Type Cultures
K_M	Michaelis constant
L-Asn/L-ASN	L-asparagine

L-Gln/L-GLN	L-glutamine
LA Activity	L-asparaginase activity
LG Activity	L-glutaminase activity
LINCS	Linear Constraint Solver
LM	Levenberg-Marquardt
MD	Molecular Dynamics
MMFF	Merck Molecular Force Field
MS	Magnesium Sulphate
MSA	Multiple Sequence Alignment
MTCC	Microbial Type Culture Collection and Gene Bank
MTT	3-(4,5-dimethylthiazol-2-yl)-2,5-diphenyl tetrazolium bromide
NCBI	National Centre for Biotechnology Information
NN	Neural Networks
NPT	Constant-temperature, constant-pressure ensemble
NVT	Constant-temperature, constant-volume ensemble
OD	Optical Density
OFAT	One-Factor-At-A-Time
PAGE	Polyacrylamide Gel Electrophoresis
PDB	Protein Data Bank
PME	Particle Mesh Ewald
PRODRG	A server for high-throughput crystallography of protein-ligand complexes
QMEAN	Qualitative Model Energy Analysis
RAM	Random Access Memory
rDNA	Ribosomal DNA
Rg	Radius of gyration

RMSD	Root Mean Square Deviation
RMSF	Root Mean Square Fluctuation
RNA	Ribonucleic acid
RSM	Response Surface Methodology
SDS	Sodium Dodecyl Sulfate
SDS-PAGE	Sodium Dodecyl Sulfate Poly Acrylamide Gel Electrophoresis
SETTLE	A constraint algorithm which is an analytical version of the SHAKE and RATTLE algorithm for rigid water models
SPF	Spherical Polar Fourier
TEMED	Tetramethyl Ethylene Diamine
Temp	Temperature
Time	Incubation Time
TSC	Trisodium Citrate
UV	Ultra-violet
WBC	White Blood Cells

INTRODUCTION

1.1. General

Among the major human diseases, cancer is a most dangerous disease. The name alone frights utmost people. The cancer cells can hide and propagate for a decent duration unnoticed. It is a second bigger disease of the human beings after Ischaemic heart disease and stroke. The number of deaths due to cancer in developed countries is three times greater than the number (85 per 100,000 populations) in developing countries(Paymaster and Gangadharan, 1971). By 2020, a 50% further increase of cancer rate is expected as per World Cancer Report.

Incidence rate of cancer in India is about 70 per one lakh population as compared to 289 per one lakh population. In India 3, 00,000 people die every year out of 5,00,000 patients. However the actual figure today could be much higher as the incidence of cancer in India is rising and many cases do not get reported. Many patients die before diagnosis of cancer registry. The burden of cancer in India has been estimated that during the beginning of the 7th plan there were approximately 1.8 million cancer cases and this figure would increase to 2.1 million cases by the beginning of the 8th plan. At the end of last century there would be around 3 million cancer cases as result of increase in size of population and longevity alone as per the report of 8th plan, Planning Commission, New Delhi, 1989.

However large number of studies carried out all over the world to control cancer, it is still remaining an unconquerable disease in the modern times. The main reasons for not devising an effective way of checking this disease are its origin, late detection and development of resistance (http://shodhganga.inflibnet.ac.in/bitstream/10603/17681/6/1_chapter%201.pdf).

The earliest documents of Chinese and Sumerian from the third millennium B.C. Physician's thought cancer is caused by disarrangement of regulatory process and tried to correct with drugs and recorded in their manuscripts. Hippocrates in 460-377 B.C. is considered as the founder of medicine and his work contains numerous references to the cause and treatment of cancer.

15

In recent years, surgery and radiotherapy are being provided for the cancer patient which are costly and risky. Few unique biochemical features of tumor cells have been selectively exploited for therapy so far. Chemotherapy has principally been focused at finding of cytotoxic agents that are able to inhibit various mammalian cell division events(Creasey, 1981). This scheme has led to development of anticancer substances of naturally occurring compounds made by microbial cell, plant cell and more recently mammalian cells cultures. The extensive information on the discovery of the naturally occurring anti-cancer agents is well summarized in recent reviews and monographs (Aszalos, 1982; Aszalos and Berdy, 1978; Cannon, 1978; Cassady and Douros, 1980; Douros and Suffness, 1981).

Uncontrolled division of cells is defined as cancer. Cancer of white blood cells (WBC), characterized by the excessive multiplication of malignant and immature WBC (lymphoblast) in bone marrow is referred to as acute lymphoblastic leukemia (ALL). Methods of treatment of acute leukemia include chemotherapy, radiation therapy, steroids and intensive combined treatments including stem cell or bone marrow transplants. Chemotherapy is the most preferred way of treatment among the above. Various drugs that are employed for treatment of ALL includes asparaginase, dexamethasone, prednisolone, vincristine, daunorubicin, cyclophosphamide, cytarabine, etoposide, thioguanine, mercaptopurine, hydrocortisone and methotrexate.

L-asparaginase enzyme is one of the potent therapeutic agents used for treatment of acute myelocytic leukemia, acute lymphoblastic leukemia in children, acute myelomonocytic leukemia, Hodgkin disease, lymphosarcoma treatment, chronic lymphocytic leukemia, melanosarcoma, and reticulosarcoma (Stecher, et al., 1999; Verma, et al., 2007). As it has an antioxidant property (Moharam, et al., 2010), it finds application in food industry as a processing aid for the effective reduction of the acrylamide levels up to 90% in starchy fried foods with no alteration in taste and texture of final product (Hendriksen, et al., 2009). Higher rates of hypersensitivity reactions restricts the prolonged use of this anticancer enzyme (Reynolds and Taylor, 1993) and anti-asparaginase antibody development results in an anaphylactic shock or neutralization of the drug effect. To overcome the above said limitations, reformed L-asparaginases from other diversified sources, L-asparaginase laden into erythrocytes, and pegylated preparations have been considered for use recently (Thomas, et al., 2010). *Erwinia chrysanthemi* and *E. coli* L-asparaginases have been more effective in acute leukemia lymphosarcoma and lymphoblastic leukemia therapy (Graham, 2003) for long years with rare therapeutic response (Duval, et al., 2002), necessitating the need to introduce fresh L-asparaginases having analogous therapeutic properties but serologically diverse. This resulted in a persistent attempt to screen novel organisms to find strains capable of fabricating new L-asparaginase with great yield and fewer adverse properties (El-Naggar, et al., 2014). Toxicity associated with L-asparaginases is partly attributable to glutaminase activity of the same enzymes (Howard and Carpenter, 1972). In materialization of L-asparagine by L-asparagine synthetase enzyme and also in quite a lot of other metabolic pathways L-glutamine is essential (Prager and Bachynsky, 1968). These days much of the research is focused on microbial based production of L-asparaginase free of glutaminase activity. L-asparaginases are highly specific towards L-asparagine and that are insignificantly active towards L-glutamine (Hawkins, et al., 2004).

Microbes are efficient producers of L-asparaginase due to their ease of handling, ease of culture and simple extraction and purification mechanisms, aiding in large scale production (Patro, et al., 2011). Major microbial L-asparaginases are intracellular but few secrete the enzyme outside the cell (Narayana, et al., 2008). It is advantageous to have extracellular L-asparaginase because of greater accumulation of product in culture broth, and simplified downstream process (Amena, et al., 2010).

Leukemic cell's inability to synthesize L-asparagine on their own has primed L-asparaginase enzyme as a dynamic factor of chemotherapy in acute lymphoblastic leukemia (ALL) therapy (Ohnuma, et al., 1970; Wriston and Yellin, 1973). In blood vascular system L-asparaginase hydrolyses L-asparagine into L-aspartic acid and ammonia building the leukemic cells barren of essential exogenous L-asparagine, leading to protein synthesis inhibition followed by apoptosis (Lubkowski, et al., 1994; Mashburn and Wriston, 1964; Neuman and McCoy, 1956), hence making this potent against ALL.

Now-a-days salable L-asparaginases are predominantly produced from *E. coli* and *Erwinia chrysanthemi* (Godfrin and Bertrand, 2006). Apart from direct production from various bacterial and fungal strains, several attempts were also made for the production of recombinant enzyme by cloning and expression of L-asparaginase encoding genes from *Erwinia chrysanthemi* (Kotzia and Labrou, 2007) and *Erwinia carotovora* (Kotzia and Labrou, 2005) and other microbes in *E. coli*. It is found that against the cells of reverted ALL patients, the commercial anti-cancer enzyme has in vitro resistance (Klumper, et al., 1995) along with its high glutaminase activity and low substrate specificity, causing pancreatitis, liver dysfunction, coagulation anomalies causing intracranial thrombosis or hemorrhage, neurological seizures and leucopenia (Klumper, et al., 1995). L-glutaminase activity of same enzyme leading to deamination of L-glutamine to L-glutamate in blood plasma induces some toxic effects in normal cells (Capizzi and Cheng, 1981; Storti and Quaglino, 1970). This facilitates the necessity for novel and healthy L-asparaginases from innocuous microorganisms with elevated substrate affinity, amended stability, low glutaminase activity, adequate half-life and least K_M value under physiological conditions to overcome the above said challenges encountered in the recent scenario. Though we have sufficient data on the production, optimization of bioprocess and purification of enzyme (Mukherjee, et al., 2000; Baskar, et al., 2009; Baskar, et al., 2011; Gurunathan and Sahadevan, 2011), no research has been done on the molecular aspects of the enzyme. In the absence of crystal structure it is highly difficult to get the molecular information about the enzyme like its interactions with the substrates and enzyme stability.

Currently L-asparaginase purified from *E. coli* is extensively used in clinical treatment of leukemia and is available in the market with the brand name of Elspar[®] (PDB ID: 1NNS). The likely side effects with Elspar[®] includes intensive allergic reactions (hives; rash; difficulty breathing; itching; tightness in chest and swelling of the face, mouth, tongue, or lips); swelling at the injection site; redness, pain, fever; liver problems (e.g., pale stools, dark urine, loss of appetite, nausea, yellowing of the eyes or skin and unusual tiredness); pancreatitis symptoms (e.g., severe back or stomach pain with or without vomiting or nausea); neurological seizures and induction of anti-asparaginase antibodies that inactivate the anti-cancer enzyme (Heinemann and Howard, 1969; Savitri and Azmi, 2003; Verma, et al., 2007).

To conquer the toxicity allied with asparaginase preparations from existing sources there is a need for new asparaginase from a diverse source that is serologically unlike but with same therapeutic result. To find an alternative and better source of L-asparaginase there is a huge enduring attention towards screening of various organisms of various biodiversities.

Though the L-asparaginase enzyme from guinea pig serum has no L-glutaminase activity (Tower, et al., 1963), bacterial L-asparaginase exhibits its activity with L-glutamine as a substrate (Campbell, et al., 1967; Campbell and Mashburn, 1969; Howard and Carpenter, 1972; Roberts, et al., 1972; Tosa, et al., 1972; Wriston, 1971). These two activities have been studied in *Escherichia coli* preparations (Miller and Balis, 1969; Verma, et al., 2007). Upon the treatment with L-asparaginase a noticeable reduction in both intra and extracellular glutamine levels has been increased both *In vitro* (Bussolati, et al., 1995; Uggeri, et al., 1995) and *In vivo* (Ollenschläger, et al., 1988; Reinert, et al., 2006; Rudman, et al., 1971). In many tissues a severe metabolic stress is caused by glutamine starvation, is followed by the Glutamine Synthetase (GS) expression and/or activity up-regulation, which acquires glutamine from glutamate and ammonium (Lacoste, et al., 1982). Furthermore, in the same cells GS activity inhibition eliminates the resistance to asparaginase cytotoxic effects of leading to gigantic cell demise. In those cells that are less sensitive to anti-cancer enzyme drug, the effects of asparaginase are significantly enriched by inhibition of GS (Tardito, et al., 2007).

1.2 Aim of the study

All the annoying qualities associated with the enzyme drug indicates the essentiality for novel L-asparaginase from other microbial sources with a great substrate kinship, improved stability, tolerable half-life with less K_M value and truncated glutaminase activity. Though adequate evidence on molecular interactions (Erva, et al., 2016; Reddy, et al., 2016), enzyme production, bioprocess optimization and purification of drug from *E. coli* and Erwinia is open, molecular aspects of the drug from *Enterobacter aerogenes KCTC 2190/MTCC 111* are not revealed by researchers. It is tough to get the molecular information involving enzyme interactions with substrate and its conformational stability in absence of three dimensional structure of enzyme.

19

Culture conditions (incubation time, pH, temperature, agitation rate and inoculum size etc.,) and medium composition greatly influences L-asparaginase production in fermentation (Hymavathi, et al., 2009). For many decades many researchers have used statistical experimental designs in biotechnological processes for an optimization purpose (El-Naggar, et al., 2014; El-Naggar and Abdelwahed, 2014; El-Naggar, et al., 2013; El-Naggar, et al., 2013) which can be implemented on several stages, in which screening the important process parameters is the prime step and the second stage is optimization of process parameters (Nawani and Kapadnis, 2005). Less experiment numbers, search for relativity between factors, suitability for multiple factor experiments and finding of the most suitable conditions and forecast of response are the advantages associated with statistical design methodologies (Chang, et al., 2006).

This laid the platform for this current study to understand the molecular information about the enzyme from novel bacterial species namely *Enterobacter aerogenes KCTC 2190/MTCC 111*, its interactions with the substrates through docking and testing the stability of the enzyme and docked complexes under physiological conditions by molecular dynamics and simulations methods. Statistical modeling tools like MINITAB, Design Expert 7.0 and MATLAB 2009a etc., for enhanced enzyme production were evaluated to optimize media components and process conditions in submerged fermentation by novel *Enterobacter aerogenes KCTC 2190/MTCC 111*.

20

1.3 Objectives of the study

The present study has been undertaken with the following main objectives.

- ✓ To identify L-asparaginase enzyme from a new source.
- ✓ *In silico* comparison with other sources of the enzyme.
- ✓ Molecular Dynamic Simulation studies on enzyme.
- ✓ To produce L-asparaginase from *Enterobacter aerogenes KCTC 2190/MTCC 111.*
- ✓ To optimize bioprocess parameters for production of L-asparaginase from *Enterobacter aerogenes KCTC 2190/MTCC 111.*
- ✓ To purify L-asparaginase from *Enterobacter aerogenes KCTC 2190/MTCC 111.*
- ✓ To perform anti-cancer activity testing against leukemic cell lines.
- ✓ To determine the ability of acrylamide degradation by L-asparaginase.

REVIEW OF LITERATIRE

2.1 Cancer

Cancer is the current major disease around the world (Organization, 2014) which involves the abnormal growth of the cell as well as has the potential/ability to invade to other parts. Various types of cancers are reported all over the world. However depending on the ability of invasion, the types of the cancer are defined into two types benign tumor and malignant tumors (Zöller, et al., 1997). Benign tumors are not having the ability to invade whereas the malignant can (King, et al., 2003). Around 6 to 12% of cancers are caused due to the inheritance (Tischkowitz and Rosser, 2004). Cancer cells spread either through lymph or blood routes (Chambers, et al., 2002).

Nowadays, around 130 different types of cancers are found. Cancers are majorly grouped into five types; Sarcoma, Carcinoma, Myeloma and Lymphoma, Leukemia and Central nervous cancers (Kasiske, et al., 2004). As per WHO, Ischaemic heart disease and stroke are the world's biggest killers, accounting for a combined 15 million deaths in 2015. These diseases have remained the leading causes of death globally in the last 15 years. Cancer is competing with the world's first disease by which the human are suffering by its second place (Organization, 2007). Cancer is mainly caused by the defective/distinct sequence of nucleotides materializing part of a chromosome by which it is associated with increment in the cell number, alterations in mechanism of regulation in the new cell, as well as the decrement of cell death with an ability to invade the surrounding cells (Menck and Munford, 2014).

Cancer can also be found due to the genetic mutations of the functional gene, which forces the deregulation of the function leading to genetic instability with respect to the parent/original gene function which drives into the tumor genesis (Jones and Baylin, 2002). No single oncogene is responsible for all the physiological traits which are transformed (Nester, et al., 1984). These kind of genetic mutations in the cells are also known as carcinogenesis which are induced by the chemical or physical factors which are having the distinctive phase's initiation, promotion and progression. Initiation is an irreversible genetic mutation in a single gene and the

22

promotion is the increment of proliferation by which the initiated cells grow rapidly. Progression itself declares the progression by more number of genetic mutations lead to the invasive type or malignant cancers (Karin and Greten, 2005).

Lymphoma and leukemia cancers originate from the hematopoietic lineage at different stages of erythroid and myeloid differentiation which are able to spread throughout the body(Reya, et al., 2001). ALL is assorted group of lymphoid maladies which consequences from the monoclonal proliferation as well as expansion of immature lymphoid cells in the blood and bone marrow (Bennett, et al., 1989; Shapiro, et al., 1988).

Over the recent decades, the new methodologies in the treatment of the ALL leads to extended survival rates. In comparison to the children and adults survival rates are 76% and 35% respectively. In many of the cases the cause of ALL is unknown (Pui, et al., 2008; Skibola, et al., 1999). Maximum numbers of cases are registered in the dizygotic and monozygotic twins (Aricò, et al., 1999; Galetzka, et al., 2012). There is a chance of higher risk in developing of ALL with excessive chromosomal fragility (Doll and Peto, 1981; Futreal, et al., 2004). Around 20% of the ALL patients are cured by the local radiations. In the course of the treatment of ALL, the patient will suffer from the anaemia or haemorrhage. Recent studies suggest that platelet rich plasma is able to decrease the haemorrhage which is caused by the thrombocytopenia (Song, et al., 1999).

2.2 Enzymes in cancer therapy

Enzymes are naturally occurring protein molecules by the living organisms, which can able to regulate and catalyze the biochemical reactions (Smith, et al., 2000). A biocatalyst is required in the systems to carry out the reaction mechanisms. The maximum rate of reaction depends on the various factors like temperature, pH and surroundings of the living organism in which the reaction occurs (Devlin, 2006). Enzymes are the vital key components of the biological systems there by acting as the catalysts. Recent studies states that serum enzymes are playing pivot role in the diseases (Chatterjea and Shinde, 2002). Enzymes are having the wide range of applications in the area of life as well as reagents in the clinical biochemistry. Enzymes are also useful in the industries for the food processing. Enzymes show a

23

significant role in the assembly of proteins besides in the central dogma (Granner, et al., 2000).

Enzymatic reactions are faster than the conventional reactions and are more specific (Koeller and Wong, 2001; Yamada and Shimizu, 1988). As a result, screening for novel enzymes/proteins which are capable to carry out the new kind of reactions is needful. The most efficient way to find out new therapeutic value enzymes are by the screening of microbes (Ogawa and Shimizu, 1999; Shimizu, et al., 1997). Most of the cases enzymes are used in the therapy of the diseases like cancer (Sabu, 2003). The word cancer originates from the Greek physician Hippocrates during the period 460-370 BC. Microbial techniques explore the new wealth for the human in terms of metabolites. Primary metabolites are the products of the metabolism which are essential like amino acids, nucleotides, antioxidants and organic acids etc., These primary metabolites are useful in the feed and food industries (Bu" Lock, et al., 1965). Whereas secondary metabolites are the antibiotics which targets the transcription, translation, DNA replication (STRHOL, 1997) and few secondary metabolites are having the enormous range of application in the pharmaceuticals and agriculture (Demain, 1990). These include compounds with hypotensive, anti cholesterolemic, antitumor, anti-inflammatory actions, plant growth regulators and also insecticides besides environmental friendly pesticides and herbicides.

Before the microbial technology, enzymes production is well known business in the industry (Katchalski-Katzir, 1993). With the help of r-DNA technology the production of specific enzymes are increased to specific levels to meet the requirement (Backman, 1987). Pharmaceutics are the major consumers of the enzymes, thereby forming the great demand for therapeutic enzymes. As a source of the novel enzymes, the plants, terrestrial and the marine microbes are exists in the nature. However, the microbes are preferred than the plants due to the cost in the production and ease in the modification. Due to the emergence of the rDNA techniques, large amount of pharmaceutical proteins are produced in less time and the cost of production is reduced (Gambardella, 1995). Currently the market is having the $3 billion value. Few enzymes are found specific to some organs/ tissues.

Some of the enzymes contribute in the radical formation as well as in the formation of super oxides (Kellogg and Fridovich, 1975; Pratt, et al., 1990).

Molecular size is the major problem in terms of the biocatalyst which is preventing in the distribution and another important problem is the elicitation of the response. Few enzymes can degrade the amino acids which are having the commercial valve in the cancer treatment. According to a report (Jemal, et al., 2011) 7.6 million people died out of 12.7 million cancer patients. In India the rate of cancer patients increased by 11% every year. One out of 5 Indian men died due to the cancers which are caused by the tobacco products. 1 in 1000 pregnant woman is affected by the cancer (Takiar, et al., 2010). Few of the factors that cause the cancer death are tobacco, infections, diet, radiation, lack of physical activity and some environmental pollutants (Anand, et al., 2008; Biesalski, et al., 1998; Sasco, et al., 2004).

2.3 Leukemia

Leukemia is a type of blood/bone cancer by the abnormal increase of immature white blood cells which is common in the childhood (Arya, 2003; Pui, 1995). The four types of leukemia comprise acute lymphocytic leukemia, chronic myelocytic leukemia, acute myeloid leukemia, and chronic lymphocytic leukemia. Uncontrolled division of cells is defined as cancer. Cancer of white blood cells (WBC), characterized by excessive multiplication of malignant and immature WBC (lymphoblast) in bone marrow is referred to as acute lymphoblastic leukemia. Methods of treatment of acute leukemia include chemotherapy, radiation therapy, steroids, and intensive combined treatments including stem cell or bone marrow transplants (Alexanian, et al., 1969). Chemotherapy is the most preferred way of treatment among the above. Various drugs employed for treatment of ALL include asparaginase, dexamethasone, prednisolone, vincristine, daunorubicin, cyclophosphamide, cytarabine, etoposide, thioguanine, mercaptopurine, hydrocortisone and methotrexate (Mashburn and Wriston, 1964).

2.4 Asparaginase sources and side effects

a. Bacterial sources

Bacterial asparaginases are considerable therapeutic interest in addition to being engaged in the treatment of ALL. *Erwinia carotovora* and *E. coli* L-asparaginases are biochemically as well as serologically diverse, while its toxicity and antineoplastic activity is alike. An extensive study on L-asparaginase production in various bacterial sources such as *E. coli* (Cedar and Schwartz, 1968; Mashburn and Wriston, 1964), *Erwinia aroideae* (Peterson and Ciegler, 1969), *Proteus vulgaris* (Tosa, et al., 1971), *Serratia marcescens* (Boyd and Phillips, 1971), *Streptomyces griseus* (Dejong, 1972) and *Vibrio succinogenes* (Kafkewitz and Goodman, 1974) has been carried out from the past few decades. In *Pseudomonas flourescens* also L-asparaginase production has been reported (Nilolaev, et al., 1974). *Mycobacterium phlei* (Pastuszak and Szymona, 1975), *Citrobacter freundi* (Davidson, et al., 1977), *Thermus aquaticus* (Curran, et al., 1985) and *Bacillus licheniformis* (Golden and Bernlohr, 1985) are found as a good sources of L-asparaginase. In *Tetrahymena pyriformis* L-asparaginase activity is found at the stationary phase of growth (Triantafillou, et al., 1988). Marine luminous bacteria are also capable of manufacturing of L-asparaginase (Ramaiah and Chandramohan, 1992).L-asparaginase from *Thermus thermophilus* is found to be as not hydrolyzing L- glutamine (Pritsa, et al., 2001; Pritsa and Kyriakidis, 2001). *Zymomonas mobilis* (Pinheiro, et al., 2001), *Erwinia sp.* (Borkotaky and Bezbaruah, 2002), *Staphylococcus sp.* (Prakasham, et al., 2007), *Erwinia chrysanthemi* 3937 (Kotzia and Labrou, 2007), *Bacillus circulans MTCC 8574* (Story, et al., 1993), *Streptomyces longsporusflavus* (Abdel-All, et al., 1998) and *Enterobacter aerogenes* (Mukherjee, et al., 2000) have also been reported as L-asparaginase producers.

b. Yeast Sources

The asparaginase production was first informed in *Streptomyces griseus* (Dejong, 1972). *Sacharomyces cerevisiae* produces two types of asparaginases (type-I and type-II) and the interactions of both anti-cancer agents with respective to the nitrogen source utilization was studied by Jones (Jones, 1977). *Rhodotorula sp.* was reported as L-asparaginase producer by Foda et al., (Foda, et al., 1979). Later, L-asparaginase production was also reported in *Rhodosporidium toruloides*

(Ramakrishnan and Joseph, 1996) and recombinant *Pichia pastoris* (Ferrara, et al., 2006).

c. Actinomycetes sources

Sreptomyces griseus, *Sreptomyces albidoflavus* and *Nocardia sp.* have abilities to produce L-asparaginase enzyme (Dejong, 1972; Gunasekaran, et al., 1995; Mostafa and Salama, 1979). *Streptomyces longsporusflavus* was able to synthesize intracellular as well as extracellular asparaginases (Abdel, et al., 1998). *Streptomyces sp.* from fish gut, *Villorita cyprinoids* as well as *Terapon jarbua* has L-asparaginase activity (Dhevendaran and Anithakumari, 2002). *Actinomycetes strain LA-29* isolated from estuarine fishes (Sahu, et al., 2007) and several actinomycetes species such as *S. karnatakensis*, *S. venezualae*, *S. longsporusflavus* and a marine *Streptomyces sp.* PDK2 have been explored for L-asparaginase production (Narayana, et al., 2008). Khamna et al., (Khamna, et al., 2009) has done L-asparaginase production from medicinal plants. Marine actinomycetes (Basha, et al., 2009), *Streptomyces gulbargensis* (Kattimani, et al., 2009) and *Fusarium sp.* (Shrivastava, et al., 2010) were also proved to be producers of L-asparaginase enzyme.

d. Plant sources

Tamarind [*Tamarindus indica*] as well as green chillies [*Capsicum annum L.*] have appreciable quantity of L-asparaginase(Bano and Sivaramakrishnan, 1980). Oza et al., reported that *Withania somifera L.* [Ashwabandha] as a probable source of the enzyme based on high specific activity (Oza, et al., 2010). In *Lupinus arboreus* plant parts, emerging seeds were found to be L-asparaginase sources (Lough, et al., 1992). Two enzyme isoforms were attained from a bryophyte, *Sphagnum fallax* (Heeschen, et al., 1996). *Lupinus arboreus* and *Lupinus angustifolius* (Borek, et al., 1999) has been reported as L-asparaginase producers among plants. An expression, catalytic activity and purification of *Lupinus luteus* asparagine beta-amidohydrolase as well as its homologue was reported in *E. coli*(Borek, et al., 2004). *Pinus pinaster* and *Pinus radiate* roots were found to be this enzyme producer (Bell and Adams, 2004). From *Arabidopsis At3g16150A*, a K-dependent L-asparaginase type was characterized (Bruneau, et al., 2006).

e. Fungal sources

L-asparaginases have been studied in *Fusarium anguioides IF04467*, *Fusarium caucasicum IF05979, Fusarium culmorum IF06814, Fusarium oxysporum IF05942, Fusarium roseum IF08503, Fusarium solani IF05232* (Nakahama, et al., 1973) and *Aspergillus nidulans* (DRAINAS and DRAINAS, 1985). Shaffer et al., (Shaffer, et al., 1988) and Mishra (Mishra, 2006) reported the production of L- asparaginase from *Aspergillus tamari* and *Aspergillus nidulans* using different synthetic substrates. *Mucorsp.* (Mohapatra, et al., 1996) and *Cylidrocapron obtusisporum* (Raha, et al., 1990) were also reported as sources of L-asparaginase. Fungi isolated from mangrove ecosystem of Bhitarkanika were screened for L-asparaginase activity and has been reported by N. Gupta et al., (Gupta, et al., 2009). During the screening program, *Scopulariopsis FMG 133* showed good enzyme activity over *Helminthosporium sp., Scophulariopsis sp., Paecilomyces sp.* and *T. Pestalotiopsis* in a study by Theantana et al., (Theantana, et al., 2007). Soniyamby et al., Hosamani et al., and Gurunathan et al., have studied L-asparaginase in *Pencilium sp.* (Soniyamby, et al., 2011), *Fusarium equiseti* (Hosamani and Kaliwal, 2011) and *Aspergillus terreus* (Gurunathan and Sahadevan, 2011), respectively.

f. Algal sources

L-asparagine specific asparaginase was purified from marine *Chlamydomonas sp.* (Paul, 1982).

g. Animal sources

Tower et al., 1963 isolated L-asparaginase from Guinea pig serum. Sayed et al., (El-Sayed, et al., 2011) investigated that purification of L-asparaginase from chicken liver and did a comparative study of some of its biochemical and biological properties.

Recently L-asparaginase has evolved as an important enzyme in growing enzyme industry, owing to its potential use in the treatment of ALL and lymphosarcoma (Story, et al., 1993; Verma, et al., 2007) and also in food industry to prevent acrylamide formation in fried foods at high temperatures (Pedreschi, et al., 2008). L-asparaginase catalyzes the hydrolysis reaction to form L-aspartate and ammonia of amide group of side chain in L-asparagine. By treating patients with L-asparaginase, circulating plasma pools L-asparagine levels were effectively depleted in the body, resulting in the inhibition of protein synthesis followed by inhibition of DNA then RNA synthesis. It causes apoptosis of leukemic cells, thereby makes it selective against the leukemic cells without affecting the normal cells (Nandy, et al., 1997).

As some leukemic cells are unable to synthesize the asparagine synthetase enzyme, they are totally dependent on circulating extracellular L-asparagine. Currently L-asparaginase purified from *E. coli* is extensively used in clinical action of leukemia which is available in the market with the brand name of Elspar®. The possible side effects reported with Elspar® include high allergic reactions; fever; pain, redness, or swelling at the injection site; symptoms of liver problems; symptoms of pancreatitis (e.g., severe stomach or back pain with or without nausea or vomiting); neurological seizures and induction of anti-asparaginase antibodies that inactivate the anti-cancer enzyme (Heinemann and Howard, 1969; Tower, et al., 1963; Verma, et al., 2007). To diminish the toxicity related with measures of asparaginase from the present sources, there is a need for documentation of a new serologically dissimilar enzyme which has the similar therapeutic effect. To obtain an alternative besides better source of L-asparaginase, there is a huge continuing interest to screen various organisms from various biodiversities.

Although the L-asparaginase enzyme from guinea pig serum has no L-glutaminase activity (Tower, et al., 1963), bacterial L-asparaginase exhibits its activity with l-Gln as a substrate (Campbell, et al., 1967; Campbell and Mashburn, 1969; Howard and Carpenter, 1972; Roberts, et al., 1972; Wriston, 1971). These two activities have been studied in *E. coli* enzyme preparations (Campbell and Mashburn, 1969; Miller and Balis, 1969). Upon treating patients of ALL with L-asparaginase, a noticeable depletion in both intracellular and extracellular glutamine is observed both In *vitro* (Bussolati, et al., 1995; Uggeri, et al., 1995) and In *vivo* (Ollenschläger, et al., 1988; Reinert, et al., 2006; Rudman, et al., 1971). In many tissues, a severe metabolic stress is caused by glutamine starvation and is trailed by the up-regulation and/or activity of Glutaminase Synthase (GS) that obtains glutamine from ammonium and glutamate (Lacoste, et al., 1982). Treatment with the anti-tumor enzyme yields a marked intensification in GS expression as well as a stimulus of GS activity. Furthermore, in the similar cells the self-consciousness of GS activity eliminates resistance to the cytotoxic effects of asparaginase leading to enormous cell

29

death. In those cells that are below par sensitive to the anti-tumor enzyme, the effects of asparaginase are significantly enriched by GS inhibition (Tardito, et al., 2007). This laid the platform for this current study to understand the molecular information about the enzyme and its interactions with the substrates through docking and testing the stability of the enzyme and docked complexes under physiological conditions by molecular dynamics and simulations methods.

Leukemic cell's incapability to synthesize L-asparagine on their own has primed L-asparaginase enzyme as an energetic factor of chemotherapy in ALL therapy (Ohnuma, et al., 1970; Wriston and Yellin, 1973). In blood vascular system, L-asparaginase yields ammonia and L-aspartic acid from L-asparagine by hydrolyses which building the leukemic cells barren of essential exogenous L-asparagine, leading to protein synthesis inhibition followed by apoptosis (Lubkowski, et al., 1994; Neuman and McCoy, 1956), hence making this potent against ALL.

Now a day, L-asparaginases are predominantly produced from *Erwinia chrysanthemi* in addition to *E. coli* (Godfrin and Bertrand, 2006). Apart from direct production from various bacterial and fungal strains, several attempts were also made for the production of recombinant enzyme by cloning and expression of *Erwinia chrysanthemi* L-asparaginase genes (Kotzia and Labrou, 2007), *Erwinia carotovora* (Kotzia and Labrou, 2005) and other microbes in *E. coli.* It is found that against the cells of reverted ALL patients, the commercial anti-cancer enzyme has *In vitro* resistance (Klumper, et al., 1995) along with its high glutaminase activity and low substrate specificity, causing pancreatitis, liver dysfunction, coagulation anomalies causing intracranial thrombosis or hemorrhage, neurological seizures and leucopenia (Duval, et al., 2002). L-glutaminase activity of same enzyme leading to deamination of L-glutamine to L-glutamate in blood plasma induces some toxic effects in normal cells (Capizzi and Cheng, 1981; Storti and Quaglino, 1970). This facilitates the necessity for novel and healthy L-asparaginases from innocuous microorganisms with elevated substrate affinity, amended stability, low glutaminase activity, adequate half-life and least K_M value under physiological conditions to overcome the above said challenges encountered in the recent scenario. Molecular mass of different native asparaginases ranges between 140 and 150 kDa (Aghaiypour, et al., 2001), whereas for *E. coli*, it was 133–144 kDa (Kozak, et al., 2002). Many studies have been carried out to optimize culture conditions for L-asparaginase production both in batch and continuous fermentation (Mukherjee, et al., 2000; Baskar, et al., 2009; Baskar, et al., 2011; Gurunathan and Sahadevan, 2011).

In addition to chief problems like hyperglycemia, thrombosis, immune suppression and pancreatitis, minor problems like, vomiting, allergy coma, headache and lethargy were also reported in the treatment of ALL (Mitchell, et al., 1994). Most communal side effects allied with L-asparaginase treatment comprise imbalances in the development of clotting factors such as protein C, plasminogen and anti-thrombin III (Trivedi and Pitchumoni, 2005).

Acute pancreatitis is one of the main side effects intertwined in leukemia treatment (Dubinsky, et al., 2000); which bear a resemblance to drug persuaded pancreatitis in recorded subjects besides the indications allied which include abdominal, anorexia and back pain. Patients in receipt of rigorous L-asparaginase therapy are found affected by means of myocardial infarction as well as fortuitous of emerging secondary leukemia (Clavell, et al., 1986) which can be convinced as a consequence of topoisomerase engaged drugs. Difficulties related through immunodeficiency besides acute hepatic dysfunctions are the foremost adjacent possessions of L-asparaginase in leukemia therapy (Kantarjian, et al., 2000). During ALL therapy, inception of venous thrombosis in children (Sahoo and Hart, 2003) and in teenage subjects cerebral thrombotic complications are pragmatic (Andrew, et al., 1994).

Growth hormone deficiency predominantly in children besides amplified risk of thrombosis is detected in patients cured for ALL (Potenza, et al., 2004). Thromboembolic difficulties in pediatric subjects with ALL are owing to deprived guideline of thrombin in addition to pro-thrombin intensities in the blood afterward the L-asparaginase treatment. L-asparaginases are found to be accompanying with corneal toxicity in subjects given with combined chemotherapy. Blur red vision, distant body sensation, bilateral conjunctival hyperaemia, ocular pain, are found to be the regular symptoms (Sutow, et al., 1971). Myocardial ischemia in ALL patients because of L-asparaginase treatment has been informed (Saviola, et al., 2004). Smooth indications correlated to diabetes are seen owing to insulin levels are decreased besides damage of pancreatic islets (Haskell , et al., 1969).

31

Instantaneous incidence of abnormalities and parotitis in lipid metabolism are observed throughout L-asparaginase treatment (Meyer, et al., 2003). Primary information indicated that there is a reduction in serum cholesterol as well as triglycerides in many subjects (Land, et al., 1972). In subjects through T-cell lymphoblastic lymphoma, L-asparaginase accompanying hyperlipidemia with hyper viscosity is also reported (Meyer, et al., 2003). Numerous information shows that hyper triglyceridemia was observed mostly in childhood subjects besides in very rare cases in adults enduring L-asparaginase treatment. Impairments of Central Nervous System utilities along with disorientation, hallucinations, agitation, convulsions in addition coma are witnessed and reported that there is an intensification in the indications of nervous syndromes after the direction of L-asparaginase (Jürgens, et al., 1987).

Asparaginase found to stimulate hypersensitive reactions in addition to the assembly of high titers of IgG3 antibodies accompanying to a higher possibility of anaphylaxis. Numerous information clearly designate that *E. coli* L-asparaginase causes additional hypersensitive reactions after compared to *Erwinia sp.* To overcome the problem of hypersensitive reactions of enzyme, PEG coated enzyme is preferred (Rossi, et al., 2004). Most of the reported side effects of the enzyme drug have arisen from the dual substrate specificity of L-asparaginase. In the therapy of ALL, enzymatic drug L-asparaginase is used to decrease the concentration of circulatory asparagine, and by that it ceases the growth of tumor cells. Asparagine and glutamine differ structurally with only one methyl group, and hence L-asparaginases will have dual substrate specificity and thus results in decreasing concentrations of both amino acids in the body (Bussolati, et al., 1995). Due to the glutaminase side activity of L-asparaginase, patients may suffer from various side effects such as immune suppression, leucopenia, thromboembolysis, acute pancreatitis, hyperglycaemia as well as neurological seizures. Many sources produce the enzyme L-asparaginase but enzymes from bacterial sources like *E. coli* and *Pectobacterium carotovorum* show the lowest toxicity related to glutaminase side activity, and because of this property there was a great interest in biomedicine and biotechnology application (Krasotkina, et al., 2004). For the first time, Manna *et al.*, (Manna, et al., 1995), reported in the side effects involved in L-asparaginase therapy was due to contamination by glutaminase of L-asparaginases from many of the sources. Glutaminase activity was found to be observed in majority of the sources that produce enzyme L-asparaginases (Imada, et al., 1973). Relatively, a very small amount of glutaminase activity was observed for the enzyme sourced from a bacterium *Pectobacterium carotovorum* among all that produced the enzyme. Hence, for better clinical studies of ALL therapy, glutaminase free L-asparaginase offers more advantages (Ramya, et al., 2012).

Attempts to minimize glutaminase side action of enzyme L-asparaginase, various attempts were made to produce glutaminase free L-asparaginase, which is regarded as a major reason for many of the side effects in the treatment of ALL (Kumar, et al., 2011). Initial attempt for the production/fabrication of glutaminase free L-asparaginase produced from *Pseudomonas* sp. was made where purified enzyme was obtained with an overall recovery of 27.2 % (Nilolaev, et al., 1974). It was reported that substitution of glutamine as well as asparagine amino acids in domicile of active site residues Glu-63 as well as Ser-254 might decrease the glutaminase activity of L-asparaginase sourced from *Erwinia chrysanthemi* (Kotzia and Labrou, 2007). Another effort was made to utilize the enzyme obtained from *E. coli* using site-directed mutagenesis, and an enzyme with reduced glutaminase activity was developed by selected replacements of amino acids at the vicinity of the active site. The authors confirmed that replacement of Asp-248 has greater effect on glutamine turnover, more strongly than asparagine hydrolysis, and suggested that small conformational changes in the active site and relevant water molecules help in attaining reduction in glutaminase activity (Derst, et al., 2000).

2.5 L-asparaginase at molecular level

Modeling is a potent methodology to specify the atoms position in a biomolecular system to a program. In general, Cartesian and Z-matrix methods are used to specify the position. The unkonown structure can also be predicted by knowing the parameters such as bond length, torsion angle and bond angle (Gajic, et al., 2011; Lawrenz, et al., 2011; Park, et al., 2003).

In the current days, computational methodology plays an important role in the prediction of the conformational changes of the system. Due to the vast advances in the technology, one can be able to get the conformational changes till the femto second and nanosecond which are experimentally inadequate to examine. The major objective of the computational biologist is to understand the biomolecular systems and their complexes (Smith and Sansom, 1998), which are dependent on the basic principles like hydrogen bonding, electro static forces and Vander Wall forces. Multiple biological systems are under investigation due to their increment in size, complexity and its assemblies by using the computational methods (Lemieux, et al., 1980; Slater, et al., 2009; Veluraja and Rao, 1980). Biomolecular complexes are able to simulate using Molecular mechanics (MM), Monte-Carlo (MC) and Molecular Dynamics (MD) simulation (Bonneau and Baker, 2001; Durrant and McCammon, 2011; Norberg and Nilsson, 2003; Senn and Thiel, 2009; Vasudevan and Balaji, 2001).

In late 1950's Alder and Wainwright introduced the molecular dynamics method (Alder and Wainwright, 1957; Alder and Wainwright, 1959). Bovine pancreatic trypsin inhibitor was the first protein simulated in the year 1977 (McCammon, et al., 1977). MD simulations are being pivotal role to study the enzymatic reactions in the context of the protein and its complexes. The frequently used software packages for MD simulations are AMBER (Case, et al., 2005; Case, et al., 2006), CHARMM (MacKerell, et al., 1998), NAMD (Phillips, et al., 2005), GROMOS (Lins and Hünenberger, 2005; Scott, et al., 1999).

Asparaginases of bacterial origin can be typeified into type I and type II on the basis of cellular location and substrate affinity (Campbell, et al., 1967). The Type1 enzymes can hydrolyze both asparagine and glutamine and are expressed in the cytoplasm whereas for expression TypeII enzymes necessitate anaerobic environments which are expressed in the periplasm. Type II L-asparaginases are used for clinical applications as they are believed to possess neoplastic activity. Type II asparaginase enzymes shows a greater L-asparagine kinship, a chief nutrient for cancer cells (Kumar, et al., 2009; Kumar, et al., 2010). The existence of type-II L-asparaginase has been stated in several bacteria such as *Erwinia carotovora* (Howard and Carpenter, 1972), *Escherichia coli* (Cedar and Schwartz, 1968; Mashburn and Wriston, 1964), *Pseudomonas sp.* (Nilolaev, et al., 1974) and *Bacillus sp.* (Golden and Bernlohr, 1985) etc.,

However, type I cytoplasmic L-asparaginase is responsible for cell metabolism. Type I asparaginase was less valued since it has lesser affinity toward the substrate L-asparagine (Schwartz, et al., 1966). The first type I bacterial L-asparaginase from *Pyrococcus horikoshii* crystal structure was unraveled only at a resolution of 2.6 Å, whereas type II L-asparaginase structure has been studied broadly (Yao, et al., 2005). L-asparaginases from different bacterial origin with solved crystal structures along with corresponding PDB IDs are mentioned in table 2.1.

Table 2.1 L-asparaginases from different bacterial origin with PDB IDs

Enzyme	PDB Code	Source Organism	Specificity
L-asparaginase	3ECA/1NNS	*Escherichia coli* (EcA2)	L-asparagine
	1WSA	*Wolinella succinogenes* (WsA2)	L-asparagine
	1HFJ/1HFK/1HFW /1HG0/1HG1/1O7J 1JSL/1JSR	*Erwinia chrysanthemi* (ErA2)	L-asparagine/ L-glutamine
	1AGX	*Acinetobacter glutaminasificans* (AGA2)	L-asparagine/ L-glutamine
	3PGA/4PGA/1DJO /1DJP	*Pseudomonas* 7A (PGA2)	L-asparagine/ L-glutamine
	1WLS	*Pyrococcus horikoshii* (PhA1) (type I)	L-asparagine

Many researchers attempted to determine the amino acid sequences, crystal structures and the molecular interactions of L-asparaginases with substrates namely L-asparagine andL-glutamine. Type II L-asparaginase from *Wolinella succinogenes, Acinetobacter glutaminasificans and Pseudomonas* 7A were reported by Lubkowski *et al.,*(Lubkowski, et al., 1996; Lubkowski, et al., 1994; Lubkowski, et al., 1994). These were compared against *E. coli* and *Erwinia* sp. L-asparaginases. Sanches et al., studied structural aspects, complexation with substrates and mechanism of reactions of bacterial asparaginases (Sanches, et al., 2007). Ramya et al., made an attempt to reduce the glutaminase side activity associated with asparaginase from *Erwinia carotovora* using *In silico* engineering approach (Ramya, et al., 2011). A similar kind of work was done by Long et al., to improve the thermo stability and catalytic activity in *Bacillus subtilis B11-06* (Long, et al., 2016).

Yaacob *et al.,* characterized a new J15 asparaginase from *Photobacterium sp.* *strain J15,* studied the molecular interactions with both substrates and also performed the Molecular Dynamic Simulation studies for the same enzyme (Yaacob, et al., 2014). However, molecular interactions of the two FDA approved enzyme drugs (Elspar® and Erwinaze®) with substrates were not extensively studied and reported in literature. Shin et al., determined the amino acid sequence of L-asparaginase form *Enterobacter aerogenes KCTC 2190* (Shin, et al., 2012). Till date either the crystal structure of the enzyme or the molecular interactions with substrates is not reported elsewhere in literature.

2.6 Statistical optimization of asparaginase production

Even though microbial production and purification of L-asparaginase are well established, yields of the enzyme have been low (Kenari, et al., 2011). So, screening of physical and nutritional parameters and evaluation is the considerable phase in the progress of bioprocess. In this respect, studying the effect of one variable at a time which is a traditional optimization technique of bioprocess is time consuming, expensive and tedious. In contrast, statistical optimization methods are favored in general because of their advantages (Dasu and Panda, 2000; Reddy, et al., 2008) and statistical experiments shrink the error in defining the effect of variables in a reasonably priced way (Kumar, et al., 2009; Sharma and Satyanarayana, 2006). The traditional one-factor-at-a-time (OFAT) technique does not reveal the interactions between different factors though a huge number of experiments have been conducted. Response Surface Methodology (RSM) (Himabindu, et al., 2006)and Taguchi methodology (Prakasham, et al., 2007) are some of statistical methods through which above said limitation can be overcome and are increasingly being used in process optimizations.

RSM reveals the interaction effects among numerous process and response variables that can be quantified, and the tool is a commanding method for testing several factors of bioprocess and offers less number of experimental runs than OFAT method. Sometimes, quadratic polynomial generated in RSM model-building stage is incapable to symbolize a given relationship to the preferred degree of accuracy, so confirming the applicability of RSM to all modeling and optimization studies is difficult (Bas and Boyaci, 2007). The alternative techniques in this perspective include artificial neural networks (ANNs) and genetic algorithms (GAs). An ANN model mimics the learning aptitude of brain that takes a whole 'black box' methodology to model the data. It is capable to model almost all types of nonlinear functions and a past knowledge of the system dynamics is not a requisite (Soria, et al., 2004).

GAs are optimization algorithms which are unorthodox search based and help in the direct search for an elucidation to a problem by imitating part of the process of natural evolution. Through a given set of alternatives GAs perform direct random searches to find the finest choice with regard to specified criterion for goodness of fit, that are expressed as a fitness function. Use of ANNs and GAs in biochemical engineering and environmental biotechnology is well established, with applications ranging from pattern recognition in chromatographic spectra, modeling of analytical biochemistry signals, cancer research, expression profiles, to functional analyses of genomic and proteomic sequences, analyzing changes in soil microbial community composition in response to hydrocarbon pollution and bioremediation etc., (Almeida, 2002).

In contrast to attempts made at protein level modification, certain trials were made at the level of fermentation for the productions of glutaminase free L-asparaginase. One such attempt was the productions of glutaminase free L-asparaginase from the bacterial source *Pectobacterium carotovorum MTCC 1428*. Statistically grounded investigational designs were applied to obtain glutaminase free L-asparaginase by Plackett–Burman process (Kumar, et al., 2009). The conditions of physical parameters found were temperature (29.8 °C), pH of the medium (6.90), shaking (157 rpm) and inoculum size [2.61 % (v/v)]. Results were obtained using central composite design technique. Under the aforesaid optimal conditions, maximum productivity of 35.24 U/mg was obtained from the bacterial source *Pectobacterium carotovorum* (Kumar, et al., 2010). Recently, attempts were also made *In silico* studies to obtain the enzyme with reduced glutaminase activity by replacing one of the enzymes active site residues Asp-96 with alanine, which helped to achieve reduction of glutaminase activity by 30% and enhanced L-asparaginase activity by 40 %. Enzyme activity was retained as such even after mutagenesis at the active site, which was confirmed through docking studies. Various microbial sources for the production of this anti-cancer enzyme drug, culture conditions and respective

37

enzyme activities achieved were given in table 2.2 (Abdel-Fattah and Olama, 2002; Amena, et al., 2010; Arrivukkarasan, et al., 2010; Derst, et al., 2000; El-Bessoumy, et al., 2004; Geckil, et al., 2004; Geckil and Gencer, 2004; Geckil, et al., 2004; Hymavathi, et al., 2009; Khushoo, et al., 2005; Kumar, et al., 2010; Liu and Zajic, 1973; Matias da Rocha Neto, 2006; Prakasham, et al., 2007; Pritsa and Kyriakidis, 2001; Venil and Lakshmanaperumalsamy, 2009; Venil, et al., 2009; Wei and Liu, 1998). Though there are several attempts for production of L-asparaginase from diverse bacterial sources, no report was made till date on the use of *Enterobacter aerogenes KCTC 2190/MTCC 111*, optimization of culture conditions and purification of enzyme etc., providing a scope to test the potentiality of this bacterium towards the L-asparaginase research.

Table 2.2 Various bacterial sources, media, culture conditions and activity values obtained in L-asparaginase enzyme production

S.No	Producer Organism	Media Used	Culture Conditions	Type of Fermentation	Nature of Enzyme	L-asparaginase Activity
1	Erwinia aroideae NRLL-B	Lactose – 1%, Yeast extract – 1.5%	24°C, pH – 7.5, 200 rpm, 12 h	Submerged	Intracellular	4 IUml-1
2	E. coli	Yeast extract – 4%, Peptone – 2%, L-asparagine – 0.1%	37°C, 220 rpm, 12 h	Submerged	Extracellular	60.8 IUml-1
3	Thermusthermophilus	Tryptone – 0.5, Yeast extract – 0.3%, NaCl – 0.2%, Glucose – 0.1%	70°C, pH – 7	Submerged	Intracellular	0.494 IUml-1
4	Pseudomonasaeroginosa 50071	Casein hydrolysate – 3.11%, Corn steep liquor – 3.68%	37°C, pH – 7.9, inoculum – 1%, 4 days	Solid State	Extracellular	0.142 U/mg
5	Enterobacter aerogenes	Peptone 1%, yeast extract – 0.5%, NaCl – 1%, glucose 0.1%	37°C, 200 rpm, 24 h	Submerged	Extracellular	510×10³ U/mg
6	Pseudomonas aeruginosa	Peptone 1%, yeast extract – 0.5%, NaCl – 1%, glucose –0.1%	37°C, 200 rpm, 24 h	Submerged	Extracellular	210×10³ U/mg
7	Pseudomonas aeruginosa 50071	Soya bean meal – 10 g. Moisture content 100%	37°C, pH – 7.4, inoculum – 1%, 96 h	Solid State	Extracellular	17.9 U/mg
8	Enterobacter aerogenes	LB media with glucose – 0.1%	37°C, 200 rpm, 24 h, inoculum – 1%	Submerged	Intracellular	113×10³ IU/mg
9	Recombinant E. coli BL 21	TB media with ampicillin 100 µg/ml	37°C, 220 rpm, pH – 7.2, 24 h	Submerged	Intracellular	22 IUml-1
10	Zymomonasmobilis cp4	Molasses – 10%,	30°C, inoculum –	Submerged	Intracellular	16.55 IUml-1

		Yeast extract – 0.2%	10%, 21 h	Submerged	Extracellular	
11	*Staphylococcus sp.6A*	Ammonium chloride-1% and glucose – 0.75%	39°C, pH – 7.5, 100 rpm, 12 h, inoculum – 3%	Submerged	Extracellular	55.6 IUml-1
12	*Serratiamarcescens*	Rice bran – 10 g, L-asparagine broth medium – 40 ml	pH – 7, inoculum – 1 ml, 36 h	Solid State	Extracellular	0.0798 U/mg
13	*Pectobacteriumcarotovorum*	Yeast extract – 2.08 %, Tryptone – 0.5 %, Monosodium glutamate – 9.89%, L-asparagine –1%, Galactose – 0.9 %	30°C, pH –6, inoculum – 5%, 120 rpm	Submerged	Intracellular	3.25 IUml-1
14	*Bacillus circulans MTCC 8574*	Red gram husk – 5 g, Glucose – 1.17 g, L-aAsparagine – aAsparagine – 1.24%.Moisture – 99.5%	36.3°C, inoculum – 2.8 ml	Solid State	Extracellular	2.322 U/mg
15	*Serratiamarcescens*	Sucrose – 1.25%, Peptone – 0.45%, L-asparagine –0.1%	pH – 7, inoculum – 1%, 51 h	Submerged	Extracellular	256 IUml-1
16	*Zymomonasmobilis strain CP4*	Glucose – 0.2%,L-asparagine – 0.4% along with yeast extract, peptone	30°C, inoculum – 2%, rpm – 120	Submerged	Intracellular /Extracellular	14.56 IUml-1
17	*Streptomyces gulbargensis*	Maltose – 0.5%, L-asparagine – 0.5%, Soya bean meal flour – 0.75%	45°C, pH – 8.5, inoculum – 1%, 200 rpm	Submerged	Extracellular	23.9 IUml-1

MATERIALS AND METHODS

3.1 *In silico* studies on L-asparaginase

3.1.1 Screening of organism

Unavailability of crystal structure for L-asparaginase from *Enterobacter aerogenes KCTC 2190* in any biological databases gave the scope to work on this bacterial enzyme. Further there are no *in vitro* studies on this organism towards the production and optimization. Moreover this enzyme of interest belongs to type I asparaginases and the available FDA approved drugs for ALL treatment falls into type II asparaginases. Till date no drug is available from type I asparaginases for therapy of ALL.

Hence the following approach was used for screening of organism (Fig. 3.1).

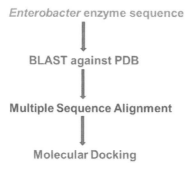

Fig. 3.1 Approach towards Screening of Organism

a. Basic Local Alignment Search Tool (BLAST)

BLAST finds regions of similarity between biological sequences. The program compares nucleotide or protein sequences to sequence databases and calculates the statistical significance of matches. BLAST can be used to infer functional and evolutionary relationships between sequences as well as help identify members of gene families. Protein-protein blast (blast p) search was performed for

Enterobacter aerogenes KCTC 2190 enzyme sequence retrieved from NCBI (ID: AEG99242.1) to identify the most similar sequences whose crystal structures were experimentally determined.

b. Multiple Sequence Alignment (MSA) by CLUSTAL Omega

Clustal Omega is a multiple sequence alignment program for proteins. It produces biologically meaningful multiple sequence alignments of divergent sequences. Evolutionary relationships can be seen via viewing Cladograms or Phylograms. This tool uses seeded guide trees and HMM profile-profile techniques to generate alignments between three or more sequences. Four type I asparaginase sequences with high percentage of sequence identity (>70%) against the *Enterobacter aerogenes KCTC 2190* L-asparaginase resulted from blastp were considered for MSA analysis to know the sequence conservation and the evolutionary relationship among them.

c. Preparation of ligands and receptor

Ligand molecules L-asparagine (L-Asn) [$C_4H_8N_2O_3$] and L-glutamine (L-Gln)[$C_5H_{10}N_2O_3$] whose molecular masses are 132.12g/mol and 146.14g/mol were retrieved from Zinc database with ID numbers 1532556 and 1532526 respectively (Fig. 3.2). Crystal structures of type I L-asparaginases (PDB IDs: 2P2D, 2OCD, 3NTX and 2HIM with more than 70% similarity with *Enterobacter aerogenes KCTC 2190*) were retrieved from PDB. As there is the lack of crystal structures for *Enterobacter aerogenes KCTC 2190* L-asparaginase, homology modelling approach was used to investigate the tertiary structures. Hypothetical configuration of *Enterobacter aerogenes KCTC 2190* L-asparaginase enzyme was obtained by MODELLER 9.13 tool using amino acid sequence.

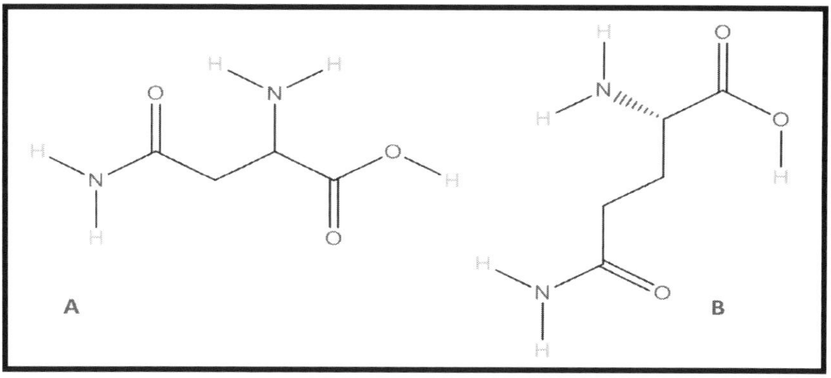

Fig. 3.2 Chemical structures of ligands   A) L-asparagine   B) L-glutamine

Further the modeled enzyme was validated using Ramachandran Plot analysis by Rampage (Lovell, et al., 2003), followed by determination of QMEAN 6 score (Benkert, et al., 2008), DFire energy value (Zhou and Zhou, 2002) using Swiss-Model server and ERRAT 2.0 (Colovos and Yeates, 1993) tools to verify the steriochemical quality of the model by analyzing the phi (ø) and psi (ψ) torsion angles, estimation of local quality of the modeled enzyme, assessment of non-bonded atomic interactions and for appraising the growth of crystallographic model construction and refinement respectively. Then all the ligands and receptors were subjected for energy minimization using the MMFF (*Merck Molecular Force Field*) (Halgren, 1996; Halgren, 1996; Halgren, 1996; Halgren, 1999; Halgren, 1999; Halgren and Nachbar, 1996) of VLifeMDS v 4.3 that works based on MM3 force fields until reaching global minima.

d. Molecular docking

Docking using Hex 8.0.0

Hex is a rigid-body docking tool for use with large molecules such as DNA and proteins. By assuming ligand conformation as rigid Hex computes protein ligand docking using Spherical Polar Fourier (SPF) relationships to accumulate the calculations(Sridhar, et al., 2005). Global docking score can be attained as a function of the six degrees of freedom in rigid docking, by scripting expressions for the overlay of pairs of parametric functions (Ritchie, 2003; Ritchie and Kemp, 2000). Docking was performed between the type I asparaginases and ligand molecules. The docking score was obtained using the default parameters and the same was interpreted as interaction energy between the ligand and receptor.

3.1.2 *In silico* comparison with type II enzyme drugs

Based on the docking results obtained from Hex, an *In silico* comparison was done with the existing FDA approved drugs from *E. coli* and *Erwinia chrysanthemi*. Crystal structure of L-asparaginase II from *E. coli* was obtained from PDB (1NNS). A homology model was developed for *Erwinia* L-asparaginase using the sequence from Drugbank (http://www.drugbank.ca/drugs/DB08886) by MODELLER 9.1. Modeled structure was subjected for quality check using the same approach used in case of *Enterobacter aerogenes KCTC 2190* modeled enzyme. Both of these receptors were energy minimized as per the procedure mentioned above.

Both the ligand substrates were docked against type II asparaginases using HEX using the default parameters of the tool and the docking energies were recorded.

a. Docking using iGEMDOCK

iGEMDOCK v 2.1 is a graphical atmosphere for identifying pharmacological interactions and virtual screening that are beneficial for pinpointing lead compounds and understanding mechanism of ligand binding against a therapeutic target. iGEMDOCK, a flexible docking tool can be used for the docking, virtual screening and post-screening analysis. The post analysis tools works by using K means and hierarchal clustering methods (Hsu, et al., 2011).

Interactive interface was provided initially for preparing target protein's binding site and compound library screening in GEMDOCK. The complete modeled structure of receptors were uploaded in the "Prepare Binding Site" and the "By current file" choice was kept because the uncut surface will be checked for binding, instead of specifying a single cavity. Then the in-house docking tool GEMDOCK docks the compounds from library into receptor binding site. Protein-compound interaction profiles were generated and analyzed by post-analysis tools. iGEMDOCK

44

finally ranks and visualizes the compound based on energy based scoring function and pharmacological interactions (Kaladhar, 2012). During docking process GA parameters were set as population size of 300, generations of 80 and solutions number as 10. Stable docking (slow) was performed with the docking scoring function as GEMDOCK scoring function and 1.00 was set to ligand hydrophobic and electrostatic preferences. Low energy profile indicates the stable system and it represents the likely binding interaction.

b. Docking using PatchDock & FireDock

In order to verify the results obtained by Hex and iGEMDOCK, another protein-protein docking was performed by PatchDock server by submitting the structures to web server (Schneidman-Duhovny, et al., 2005) that works based on shape complementarity principles and again the outcomes were refined with FireDock server (Andrusier, et al., 2007; Mashiach, et al., 2008) that reshuffles the interface side chains and amends the molecule's relative orientation. Analysis of ligand binding interactions and docking viability was established on Fire Dock scores and visualizations with Pymol. Docking parameters were set as 30° as rotation angle, 30 as number of placements and 10 as ligand wise results. Intel Core i7 7470 CPU @ 3.40 GHz of DELL origin, with 8 GB DDR RAM under Windows 8.1 OS was used to perform all the docking runs.

3.1.3 Molecular Dynamics (MD) and Simulations

MD simulations were executed for the apo enzymes and the docked complexes gained from molecular docking to ratify the stability in dynamic system which is as follows.

1. Complex 1 = 1NNS+L-Asn
2. Complex 2 = 1NNS+L-Gln
3. Complex 3= Erwinaze®+L-Asn
4. Complex 4 = Erwinaze®+L-Gln
5. Complex 5 = *Enterobacter aerogenes KCTC 2190* L-asparaginase+L-Asn and
6. Complex 6= *Enterobacter aerogenes KCTC 2190* L-asparaginase+L-Asn

45

Generating the L-asparagine and L-glutamine topologies using PRODRG server is the early step in MD simulations (SchuÈttelkopf and Van Aalten, 2004). After defining ligand topologies, MD simulation for apo enzymes and docked complexes were carried using GROMACS 4.6.5 program package under Ubuntu 14.04 operating system. Steepest algorithm with OPLS force field (Lindahl, et al., 2001)was used for energy minimization for apostate *E. coli* enzyme, *Enterobacter aerogenes KCTC 2190* enzyme, complex 1, complex 2, complex 5 and complex 6 dismissing when the maximum force is found lesser than 1000 KJ mol^{-1} nm^{-1}. Whereas for apostate *Erwinia chrysanthemi* enzyme, complex 3 and complex 4 Steepest algorithm using the GROMOS 96 43a1 (Scott, et al., 1999; van Gunsteren, et al., 1996) was used for energy minimization, dismissing when the maximum force was found lesser than 1000 KJ mol^{-1} nm^{-1}.

To provide an aqueous environment in a cubic box system with 1.0 nm size and at least 2.0 nm in between any two periodic protein images, all the molecules were solvated. With the addition of six Sodium ions the system was neutralized and in all the directions periodic boundary conditions were employed. The cubic interpolation order in Particle Mesh Ewald (PME) simulation method is 4.0 and the grid spacing for Fast Fourier Transform (FFT) is 16×10^{-2}. In the neighbor searching method the short range neighbor list cutoff of 1Å is taken commonly for electrostatic interactions and Van der Waal interactions. The SETTLE (Miyamoto and Kollman, 1992)and LINCS(Hess, et al., 1997) algorithms were applied to restrain geometry of water molecules and entire bond lengths respectively in the system.

For all the molecules two equilibration phases namely NVT and NPT ensembles were applied for a period of 100 ps, with a constant temperature of 300°K, coupling constant of 0.1 ps and with constant pressure of 1 bar, coupling constant of 2 ps respectively. Modified Berendsen thermostat coupling scheme algorithm was employed for both ensembles of equilibration. Once the system equilibration with constant temperature and pressure is done, to perform structural and energy analyses, 30 ns production MD run was performed. Run trajectories were obtained and quality assurance of all the molecules was done with GROMACS utilities namely g_energy, g_rms, g_rmsf and g_gyrate. Hydrogen bonding study was done by g_hbond. Using Xmgrace tool the entire trajectory results were analyzed.

3.2 *In vitro* studies on L-asparaginase

3.2.1 Chemicals

All the analytical grade chemicals used in this study were from Hi-Media (India) except Nessler's reagent acquired from Finar (India).

3.2.2 Microorganism

The bacterial strain used in this work was *Enterobacter aerogenes MTCC 111* (equivalent to KCTC 2190), attained from Microbial Type Culture Collection and Gene Bank (MTCC), Chandigarh, India.

3.2.3 Inoculum Culture

Culturing of inoculum was done in growth media (table 3.1). The strain was maintained on slants and stored at 4°C. Stock culture was transferred regularly to fresh agar slant after every 4 weeks.

Table 3.1 Growth media for inoculum culture

S. No	Component	Quantity
1	Beef extract	1 g
2	Yeast extract	2 g
3	Peptone	5 g
4	NaCl	5 g
5	Agar	15 g
6	Distilled water	1 litre

3.2.4 Qualitative assay for L-asparaginase production

The bacterium was spread on media having L-asparagine along with phenol red (few drops) as indicator. The inoculated plates were subjected for overnight incubation at 37°C. Microbial culture displaying red color around colony confirms the production of L-asparaginase. Release of ammonia due to the ruin of L-asparagine increases the pH which is detected by phenol red indicator that shifts media color from yellow to pink. Modified M-9 agar medium was used for this purpose (table 3.2).

Table 3.2Modified M-9 agar medium

S. No	Component	Quantity
1	KH_2PO_4	3 g
2	$Na_2HPO_4.2H_2O$	6 g
3	NaCl	5 g
4	$MgSO_4.7H_2O$	0.5 g
5	L-asparagine	5 g
6	20% glucose stock	10 ml
7	$CaCl_2.2H_2O$	0.014 g
8	Agar	20 g
9	Distilled water	1 litre

2.5% phenol red stock was made in ethanol with pH 7.0. 0.3 ml of stock was supplemented to 0.1 liters of modified M-9 medium. Bacterial colonies surrounded with pink zone authorize the L-asparaginase production.

Confirmation of qualitative assay of L-asparaginase was performed using a rapid assay methodology (Gulati, et al., 1997) by the addition of 0.2 ml of sample to 10 ml of water having Asparagine and phenol red indicator (0.001%). L-asparagine break down turns the mixture basic because of ammonia release, altering the mixture color from yellow to pink.

3.2.5 Crude Enzyme Preparation

10 ml of fermented broth is centrifuged for 10 min at 10,000 rpm and 5°C. Resultant cell pellet is dissolved in 15 ml of 0.05 M Tris-HCl buffer (pH 8.6), homogenized and double filtered by Whatman #1 filter paper. The filtrate is considered as crude enzyme.

3.2.6 Analytical methods

a. Quantitative Assay of enzyme activity

L-asparaginase activity (LA Activity) and L-glutaminase activity (LG Activity) of the crude enzyme was analyzed by quantifying ammonia formation by spectroscopy. Nesslerization is universally used technique for assessment of L-asparaginase activity. The extent of liberated ammonia during L-asparagine hydrolysis was measured spectrophotometrically (JASCOV 600) at 480 nm by Nessler's reagent. One unit (IU) of L-asparaginase activity is defined as the quantity of enzyme which liberates 1 μmol of ammonia per minute in the standard assay conditions (Mashburn and Wriston, 1964). Ammonia liberation was estimated using an Ammonium sulfate standard plot (APPENDIX-B).

b. Quantification of total protein content

Using Bovine Serum Albumin (BSA) as the standard, analysis of the total protein content of the crude enzyme was done as per the protocol of Lowryet al., (Lowry, et al., 1951) (APPENDIX-C).

c. Microbial growth profile measurement

Growth profile of microbial culture was recorded by quantifying the Optical Density (OD) of cell suspension at 600 nm at a fixed interval of 6 h using JASCOV 600 spectrophotometer.

3.2.7 Production and optimization of L-asparaginase enzyme

a. Primary inoculum preparation and production conditions

Primary inoculum was prepared by adding pure culture of *Enterobacter aerogenes MTCC 111* in 50 ml of Luria - Bertani sterile medium and flask was kept at 30°C for 24 h in a rotary shaking incubator (Remi CIS 24 BL) at 200 rpm.

b. Optimization of L-asparaginase production by one-factor-at-a-time method

Optimization of all the process variables influencing the production of enzyme was done by one-factor-at-a-time method. Total protein, pH, bacterial

49

growth and enzyme activity profiles were recorded as per the procedure mentioned below in all instances.

Influence of incubation time

In order to validate the batch time, 0.5% (v/v) inoculum (A 600 = 0.6 to 0.8) of primary inoculum was added to 50 ml of production medium and allowed for incubation with shaking at 200 rpm and at 30 °C. L-asparaginase production was performed in production medium (Mukherjee et al., 2000) as mentioned in table 3.3.

Table 3.3 L-asparaginase production media for *Enterobacter aerogenes MTCC 111*

S. No	Component	Quantity
1	Trisodium citrate	0.75%
2	K_2HPO_4	0.0125%
3	$(NH_4)_2HPO_4$	0.2%
4	$FeSO_4.7H_2O$	0.002%
5	$MgSO_4.7H_2O$	0.002%
6	Yeast extract	0.15%
7	$CaCl_2.2H_2O$	0.014 g
8	L-asparagine	1%
9	Distilled water	1 litre

L-asparagine was added as an inducer for the enzyme production. Incubation was carried out for 54 h. For every 6 h of time frame pH and bacterial growth profiles were maintained using the fermented culture, whereas total protein and enzyme activity profiles were recorded using the crude enzyme sample. Fermented broth was centrifuged for 10 min at 10000 rpm (Remi C-30 BL) and *Enterobacter aerogenes MTCC 111*cells obtained were dissolved in 0.05 M Tris-HCl buffer (pH 7.4). The cells were homogenized by sonication in Tris-HCl buffer and cell debris was taken off by filtering through Whatman#1 filter paper. This preparation served as crude enzyme sample. All the experimentations were performed thrice and the mean values with standard deviations were calculated.

Influence of pH and temperature

Sterile production media of 50 ml (table 3.3) in 250 ml conical flasks was used for studying the effect of pH and temperature. First initial pH was varied from 5 to 9 to determine the optimum pH and using the optimum pH value, temperature was varied from 20°C to 35°C with an agitation rate of 200 rpm in both cases to determine their autonomous effect on production of L-asparaginase.

Influence of carbon sources

Sterile production media (50 ml) was prepared in 250 ml conical flasks. The influence of different carbon sources [trisodium citrate ($Na_3C_6H_5O_7$), glucose ($C_6H_{12}O_6$), starch ($C_6H_{10}O_5$) and sucrose ($C_{12}H_{22}O_{11}$)] at 0.75% (w/v) was studied along with $(NH_4)_2HPO_4$,0.2%; K_2HPO_4,0.0125%; $FeSO_4.7H_2O$, 0.002%; $MgSO_4.7H_2O$,0.02%, yeast extract,0.15% and 1% L-asparagine at optimum pH and temperature values from previous experiments. The finest carbon source was optimized for L-asparaginase manufacturing using same media containing above components. 0.5% (v/v) of inoculum was added to the media and was incubated at 200 rpm.

Influence of L-asparagine

The effect of L-asparagine as an inducer for enzyme drug of interest in the present study was determined by adding different inducer concentrations (1%, 3%, 5% and 7%) in the production medium. L-asparaginase activity was analyzed by standard L-asparaginase assay (Mashburn and Wriston, 1964). Total protein content, pH and microbial growth profiles were also recorded for every 6 h.

Influence of nitrogen sources

Influence of diverse nitrogen sources on L-asparaginase activity was studied using yeast extract, beef extract, ammonium chloride and casein at 0.15% concentrations in the production medium. Asparaginase activity was analyzed by standard L-asparaginase assay. Further concentration of nitrogen source was optimized for L-asparaginase production using same media with the best nitrogen source and 0.5% (v/v) of inoculum. The same is incubated at 200 rpm at optimum pH and temperatures.

Influence of inoculum size

In order to find out the size (%) of inoculum on the level of asparaginase making, diverse inoculum sizes (0.5, 1.0, 1.5, 2 v/v) were used in 250 ml conical flask having 50 ml of production medium. Flasks were incubated at optimized conditions with shaking at 200 rpm. Profiles for total protein, pH, growth of microbial cells and enzyme activity were analyzed as per the standard protocols.

c. Statistical optimization of L-asparaginase production

Medium optimization for maximized production of L-asparaginase by *Enterobacter aerogenes MTCC 111* was done in two phases.

Plackett-Burman design for significant medium components screening

To screen the significant culture variables influencing the production of L-asparaginase, Plackett-Burman experimental design was used (Plackett and Burman, 1946). The total number of parameters considered for screening is fourteen, *viz.*, Time, Temperature (Temp), pH, rpm, Dissolved Oxygen (DO), Inoculum size (Inn Size), Inoculum age (Inn Age), Trisodium Citrate (TSC), Di Ammonium Hydrogen Phosphate (DAHP), Magnesium Sulphate (MS), Ferrous Sulphate (FS), Di Potassium Hydrogen Phosphate (DPHP), Ammonium Chloride (AC) and L-asparagine (L-ASN). All independent variables were designated at two levels, low and high, and are indicated by (-1) and (+1), respectively.

Names of the process variables, codes and their actual levels are presented in table 3.4. Whereas table 3.5 describes the experimental design used in screening process. Production of L-asparaginase was done and enzymatic assay was performed in triplicates and the response was the average of three. As per Plackett-Burman design, a sum of 20 experiments was carried out. Using Student's t-test, significance of every process variable was determined.

Table 3.4 Variables used in Plackett-Burman design

Variables	Code Levels	
	-1	+1
Time(h)	10	40
Temp	25	40
pH	6	10
Rpm	100	200
DO	1	3
Inn Size (%)	0.25	1
Inn Age (h)	24	72
TSC	0.5	1
DAHP	0.1	1
MS	0.01	0.05
FS	0.001	0.005
DPHP	0.01	0.05
AC	0.5	1
L-ASN	0.5	2

Table 3.5 Plackett-Burman design table used for screening of variables

S.No	Time (h)	Temp (°C)	pH	rpm	DO	Inn Size (ml)	Inn Age (h)	TSC (%)	DAHP (%)	MS (%)	FS (%)	DPHP (%)	AC (%)	L-ASN (%)
1	+1	+1	+1	-1	+1	-1	-1	-1	-1	+1	-1	-1	-1	-1
2	-1	+1	-1	-1	+1	+1	+1	-1	-1	+1	-1	-1	-1	+1
3	+1	+1	+1	-1	+1	+1	+1	+1	+1	-1	-1	+1	+1	+1
4	-1	+1	-1	+1	+1	-1	-1	+1	+1	-1	-1	+1	-1	-1
5	-1	-1	-1	+1	+1	-1	+1	+1	+1	-1	+1	+1	-1	-1
6	-1	+1	+1	+1	+1	-1	+1	+1	+1	+1	-1	+1	+1	+1
7	+1	+1	-1	-1	-1	-1	-1	+1	-1	+1	+1	+1	+1	+1
8	-1	-1	-1	-1	-1	+1	+1	+1	+1	-1	+1	-1	+1	+1
9	+1	+1	+1	+1	-1	+1	+1	-1	+1	+1	-1	-1	-1	-1
10	-1	-1	-1	-1	-1	+1	+1	-1	+1	+1	-1	-1	-1	+1
11	+1	+1	+1	+1	-1	-1	+1	+1	-1	-1	+1	+1	+1	-1
12	+1	-1	-1	-1	-1	+1	-1	+1	-1	-1	+1	-1	+1	+1
13	+1	-1	+1	+1	-1	-1	-1	-1	+1	+1	+1	-1	+1	+1
14	+1	-1	-1	-1	+1	-1	+1	+1	-1	+1	+1	+1	+1	-1
15	-1	-1	-1	+1	+1	-1	+1	-1	+1	-1	-1	+1	-1	-1
16	+1	-1	+1	+1	+1	+1	-1	-1	+1	+1	-1	-1	+1	-1
17	-1	-1	-1	-1	-1	-1	-1	+1	-1	-1	+1	-1	-1	-1
18	+1	-1	-1	+1	+1	+1	+1	+1	-1	-1	+1	+1	-1	+1
19	-1	+1	-1	+1	+1	+1	-1	-1	-1	-1	+1	-1	+1	-1
20	-1	+1	+1	-1	-1	+1	+1	+1	+1	-1	-1	+1	+1	-1

54

The experimental design of Response Surface Methodology (RSM) involved in the selection of every independent process parameter consists of three levels. Among all the physical variables tested, three variables (Temperature, pH, and rpm) were found to have major impact on enzyme activity based on Plackett-Burman screening. The response functions of interest were L-asparaginase and L-glutaminase enzyme activities. The function was estimated by a second degree polynomial of quadratic and interaction effects using the method of least squares (Rajulapati, et al., 2011). The levels of Temperature (A), pH (B) and rpm (C) were considered for studies using the full factorial Face Centered Central Composite Design (FCCCD) as it includes duplication of the medial point. Actual levels of coded factors were shown in table 3.6which was determined based on the results obtained in experiments described in Plackett-Burman method. The statistical software package Design-Expert7® (Stat-Ease Inc, USA) was used for analysis of the investigational design. A sum of 20 experimentations was generated (Table 3.7) with 14 being the blends of the definite level of the investigational parameters and the remaining 6 were duplications at the central points, which were performed to establish the curvature and to balance for the lack of fit values which specify the model significance.

Table 3.6 Variables used in RSM experimental design

Variables	Codes	Code Levels		
		-1	0	+1
Temp	A	25	37.5	40
pH	B	6	8	10
Rpm	C	100	150	200

Analysis of Variance (ANOVA) was performed on anti-leukemic enzyme production data to recognize the effects of individual variables. The mathematical connection between autonomous variables and objective function (L-asparaginase and L-glutaminase production) was deliberated by the second order polynomial equation,

$$Y_1 = B_0 + \Sigma B_i X_i + \Sigma B_{ii} X_i^2 + \Sigma B_{ij} X_i X_j \qquad \longrightarrow \qquad \text{(1)}$$

$$Y_2 = B_0 + \Sigma B_i X_i + \Sigma B_{ii} X_i^2 + \Sigma B_{ij} X_i X_j \qquad \longrightarrow \qquad \text{(2)}$$

in which

Y_1- predicted response for L-asparaginase activity

Y_2- predicted response for L-glutaminase activity

B_0- intercept term

Bi - linear effect

Bii- squared effect and

Bij- interaction effect.

Equations (1) and (2) can be used to approximate the linear, quadratic and interactive influence of autonomous process parameters on the final response. Consequential 3D surface plots designate the objective function on Z-axis with X and Y-axes demonstrating the two autonomous process parameters.

Table 3.7 RSM design table for optimization of L-asparaginase and
L-glutaminase activities

Standard Order	Temp (°C)	Rpm	pH	L-asparaginase Activity (IUml⁻¹)	L-glutaminase Activity (IUml⁻¹)
1	-1	-1	-1		
2	+1	-1	-1		
3	-1	+1	-1		
4	+1	+1	-1		
5	-1	-1	+1		
6	+1	-1	+1		
7	-1	+1	+1		
8	+1	+1	+1		
9	-1	0	0		
10	+1	0	0		
11	0	-1	0		
12	0	+1	0		
13	0	0	-1		
14	0	0	+1		
15	0	0	0		
16	0	0	0		
17	0	0	0		
18	0	0	0		
19	0	0	0		
20	0	0	0		

Modeling using Artificial Neural Networks (ANNs)

ANN models imitate the role of a biological network, made up of neurons and are applied to decipher composite functions in diverse applications. Simple synchronous processing elements are included in NN which are motivated by the biological nerve systems. Neuron is the basic unit of ANN and they are linked to one another by called synapses, and a weight factor is allied with every synapse (Zhang and Friedrich, 2003). Back-Propagation (BP) is one of the trendiest algorithms in ANN which is used in this present study, with one hidden layer enhanced with numerical optimization technique named Levenberg-Marquardt (LM) (Arcaklıoğlu, et al., 2004).

Process optimization by multi objective Genetic Algorithm (GA) optimization

A theoretical universal search and optimization technique called GA, copy the metaphor of natural biological evolution. GA works on a population of likely solutions implying the principle of survival of the fittest to yield sequentially superior estimations to a solution. A fresh set of estimation is produced at each generation by the process of individual selection as per their fitness level in the domain of problem and their replication using rented operators from natural genetics. The above practice directs to progression of individual populations that are well-matched for their surroundings compared to the entities from which they were generated, as similar in normal adaptation process (Rajulapati, et al., 2011).

The GA optimization begins by initializing the population of solutions P(t). The size of population was 12(4*No. of variables) and the primary population type chosen was binary. In every chromosome the evaluation function computes the fitness value; in this study, the error between finale output and present output was the fitness function. The choice of the individuals to generate the successive generation has a vital role in GA. The apparent choice begins from each individual's fitness which provides the inaccuracy between the objective and real outputs, so that smallest error generating individual has greater chance to be elected. Many methods like Geometric ranking method, Rank selection and Roulette wheel selection etc., are used for the process of the selection and Rank method was opted in the present optimization. Crossover and mutation offer the fundamental search mechanism of a

GA. Depending on preceding solutions produced, the operators build fresh results. Crossover accepts two entities and generates two novel recombinant entities, but mutation alters the individual by arbitrary adjustment in a gene to turn out a fresh solution. Application of genetic operators and their derivatives is based on chromosome depiction. Scattered option was used as crossover operator and other constraints used for reproduction and mutations are 0.8 crossover rate and constraint dependent mutations function. Other approximated parameters were forward migration direction, 0.2 migration fraction and 20 as migration interval. The ending criterion usually advises the upper limit of repetitions or verifies if the finest solution attained is acceptable. Values considered for ending criteria includes 300 as highest number of iterations (100* number of variables), unlimited time, infinite limit of fitness, 50 stall generations, unbounded limit of stall time, function tolerance and 10^{-6} nonlinear constraint tolerance. GA optimization was used for maximizing L-asparaginase activity and minimizing the L-glutaminase activity.

All the optimized values given by RSM and multi objective GA tools were validated by performing the experiments at the given process variable values and the enzyme activities were estimated.

3.2.8 Enzyme purification and quantification

All purification steps were carried out at 0 to 4°C unless otherwise indicated.

a. Ammonium sulfate precipitation

Ammonium sulfate was added to the clear supernatant obtained after ultra-sonication with uninterrupted rotation and allowed the same for overnight incubation. With the precipitated fraction at 70% saturation highest L-asparaginase activity was perceived. By centrifugation for 10 min at 6000 rpm, precipitate was collected and dissolved in a marginal volume of 10 mM Tris–HCl buffer (pH 7.0). Against the same buffer this sample was dialyzed for 24 h.

b. DEAE cellulose chromatography

The enzyme mixture after dialysis was poured on to a column packed with di ethyl amino ethyl (DEAE) cellulose. The column was pre-equilibrated at 1 ml/min flow rate using10 mM Tris–HCl (pH 7.0) before sample loading. The chromatography column was washed using twice the volume of column using same buffer and using a NaCl (25–150 mM) as linear gradient in 10 mM Tris–HCl (pH 7.0) and the adsorbed protein was eluted. Fraction with L-asparaginase activity attained after this step of chromatography was further used in the next step of purification.

c. Sephadex G-75 chromatography

On to the pre-equilibrated Sephadex G-75 column with 10 mM Phosphate buffer with pH 7.0 active sample was loaded which was obtained from the above purification phase. At 0.2 ml/min flow rate elution was carried out with the same buffer. The collective active fractions were stored at 4°C.

d. Molecular weight determination

In a 3mm thick slab gel of 6% acrylamide in Tris-glycine buffer of pH 8.2 having 0.1% Sodium Dodecyl Sulfate (SDS), polyacrylamide gel electrophoresis (SDS-PAGE) was performed. With Coomassie brilliant blue (CBB) R-250 the gels were stained and destained (Stegemann, 1979).

3.2.9 Anti-Cancer testing of L-asparaginase on leukemic cell lines

Viability of HL-60 cells was determined by3-(4,5-dimethylthiazol-2-yl)-2,5-diphenyl tetrazolium bromide (MTT). In a 96-microwell plate, at 1×10^5 cells/well concentration cells were seeded and were permitted to adhere for 24h. Through a sterile 0.2μm syringe filter sample solution is filtered before adding to the cells. The purified L-asparaginase enzyme is supplemented at different concentrations and incubated for 72h. Controls include culture medium, HL-60 cells, but no L-asparaginase. After the completion of 72h incubation period, the medium was separated and 100μl of MTT (0.5mg/ml) was mixed and incubated for 4 h, as described previously. In 100μl of 10% SDS in 0.01N HCl crystals were solubilized and allowed for overnight incubation. At 550nm, using a micro plate reader (Thermo

Scientific™ Remel™ ELx800 Automated Microplate Reader), the absorbance was measured and as the percentage of viable cells with respect to the control cells the results were expressed.

Percentage of cell viability was measured as

$$\% \text{ Cell Viability} = \frac{\text{Mean absorbance of the sample}}{\text{Mean absorbance of the control}} \times 100$$

3.2.10 Acrylamide degradation studies

Tris–HCl buffer (pH 8.5) - 2.5 ml and double distilled water of 2.5 ml were added to 5 ml of 10% acrylamide solution. To this reaction mixture *Enterobacter aerogenes KCTC 2190/MTCC 111*crudeL-asparaginaseenzyme was added and is incubated for 30 min at 45°C. Then, 20 μl of tetramethyl ethylene diamine (TEMED) along with 200 μl of ammonium persulphate (APS) (10%) were added to the above mixture. At room temperature the tubes were incubated and the solidification time was recorded (Mahajan, et al., 2012).

RESULTS AND DISCUSSION

4.1 *In silico* studies on L-asparaginase

4.1.1 Screening of organism

a. Amino acid sequence of Enterobacter aerogenes KCTC 2190 L-asparaginase

*Enterobacter aerogenes KCTC 2190*L-asparaginase enzyme sequence (Fig. 4.1) was retrieved from NCBI (ID: AEG99242.1).

```
>MQKKSIYVAYTGGTIGMQRSDHGYIPVSGHLQRQLALMPEFHRPEMPDFTIHEYAPLMDS
SDMTPEDWQHIADDIRDHYDQYDGFVILHGTDTMAFTASALSFMLENLGKPVIVTGSQIP
LAELRSDGQINLLNALYVAANYPINEVSLFFNNRLYRGNRTTKAHADGFDAFASPNLAPL
LEAGIHIRRLGTPPAPHGKGELIVHPITPQPIGVVTIYPGISADVVRNFLRQPVKALILR
SYGVGNAPQNGEFIQVLAEASQRGIVVVNLTQCMSGKVNMGGYATGNALAQAGVISGFDM
TVEATLTKLHYLLSQNLDGAAIRNAMQQNLRGELTPDE
```

Fig. 4.1 Amino acid sequence of *Enterobacter aerogenes KCTC 2190*
L-asparaginase

b. Basic Local Alignment Search Tool (BLAST)

The above sequence is submitted to protein-protein blast (blast p) tool of NCBI to identify the most similar sequences whose crystal structures were experimentally determined. A total of four type I asparaginase sequences were identified with at least 70% sequence identity against *Enterobacter aerogenes KCTC 2190* namely, PDB IDs: 2P2D-92% identity, 2HIM-91% identity, 3NTX-82% identity and 2OCD-70% identity (Fig. 4.2).

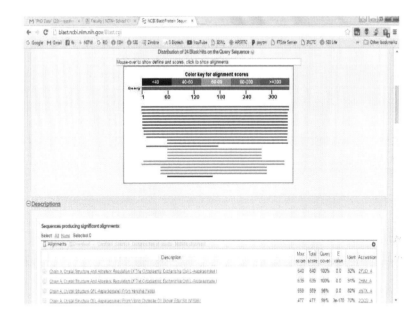

Fig. 4.2 BLAST p results of *Enterobacter aerogenes KCTC 2190* L-asparaginase

c. Multiple Sequence Alignment (MSA) by CLUSTAL Omega

MSA was performed for all the five type I asparaginase sequences which includes the *Enterobacter aerogenes KCTC 2190* L-asparaginase, 2P2D, 2HIM, 3NTX and 2OCD to produce biologically meaningful sequence alignments and to determine the evolutionary relationships among them (Fig. 4.3). CLUSTAL Omega (CLUSTAL O 1.2.1) is the tool used for this purpose.

From the analysis of results, it is determined that majority of the query sequence is conserved with respective to known crystal structures (Fig. 4.4). A cladogram has been developed and it clearly says that all the sequences are evolutionarily closely related (Fig. 4.5).

CLUSTAL O(1.2.1) multiple sequence alignment

```
2OCD:A|PDBID|CHAIN|SEQUENCE          --------------------MARNHIYIAYTGGTIGMNKSDHGYVFVASFMEKQLASMPE
3NTX:A|PDBID|CHAIN|SEQUENCE          -------------------SNANQKMSIVVAYTGGTIGMQRSDNGYIFVSGHLQRQLALMPE
EARCTC2190|PDBID|CHAIN|SEQUENCE      -------------------NQKMSIVVAYTGGTIGMQRSDHGYIFVSGHLQRQLALMPE
2HIM:A|PDBID|CHAIN|SEQUENCE          MGSSHHHHHHSSGLVPRGSHMQKMSIVVAYTGGTIGMQRSEQGYIFVSGHLQRQLALMPE
2P2D:A|PDBID|CHAIN|SEQUENCE          MGSSHHHHHHSSGLVPRGSHMQKMSIVVAYTGGTIGMQRSEQGYIFVSGHLQRQLALMPE
                                                         * :* *** *:*********** :(*:.**;**;*.;;:*** ***

2OCD:A|PDBID|CHAIN|SEQUENCE          FHRPEMPLPTIHEYDFLMDSSDMIPADWQLIADDIAANYDHYDGFVILRGTDTMAYTASA
3NTX:A|PDBID|CHAIN|SEQUENCE          FHRPEMPDFTIHEYAFLIDSSDMTPEDWQRTAHDIQQNYDLYDGFVILRGTDTMAFTASA
EARCTC2190|PDBID|CHAIN|SEQUENCE      FHRPEMPDFTIHEYAFLMDSSDMTPEDWQRIADDIRDHYDQYDGFVILRGTDTMAFTASA
2HIM:A|PDBID|CHAIN|SEQUENCE          FHRPEMPDFTIHEYTFLMDSSDMTPEDWQRIAEDIWAHVDDYDGFVILRGTDTMAYTASA
2P2D:A|PDBID|CHAIN|SEQUENCE          FHRPEMPDFTIHEYTFLMDSSDMTPEDWQRIAEDIWAHVDDYDGFVILRGTDTMAYTASA
                                     ******* .****** **:;:****** *** **:;** ;** ************;***

2OCD:A|PDBID|CHAIN|SEQUENCE          LSPMFENLGKPVTVTGSQIPLADLRSDGQANLLNALFVRAANYPINEVTLFPNNRLHRGNR
3NTX:A|PDBID|CHAIN|SEQUENCE          LSPMLENLAKPVITGSQIPLAELRSDGQTNLLNALYLAANHPANEVSLFPNNQLFRGNR
EARCTC2190|PDBID|CHAIN|SEQUENCE      LSPMLENLGNPVTVTGSQIPLALELRSDGQINLLNALVVRAANYPINEVSLFPNNRLYRGNR
2HIM:A|PDBID|CHAIN|SEQUENCE          LSPMLENLGNPVTVTGSQIPLAELRSDGQINLLNALYVRAANYPINEVTLFPNNRLYRGNR
2P2D:A|PDBID|CHAIN|SEQUENCE          LSPMLENLGNPVTVTGSQIPLAELRSDGQINLLNALYVRAANYPINEVTLFPNNRLYRGNR
                                     ****;;*** .****;:(*******;;****** ****** ;;***;;;*****;;;**** ****

2OCD:A|PDBID|CHAIN|SEQUENCE          SRMSHADGPSAFSSPNLFPILERGINIELSTNVKVDEKPSGEFKVNPITPQPIGVTIMYP
3NTX:A|PDBID|CHAIN|SEQUENCE          TTKGHADGFDIFASPNLSVLLEAGIHIRPQSSV-VSPTSNGPLIVHRITPQPIGVVTIYP
EARCTC2190|PDBID|CHAIN|SEQUENCE      TTKGHADGFDAFASPNLAPLLEAGTHIHRLGTP-PAPSGNGELIVHRITPQPIGVVTIIYP
2HIM:A|PDBID|CHAIN|SEQUENCE          TAKGHADGFDAFASPNLFPLLEAGIHIHRLNTP-PAPSGEGELIVHRITPQPIGVVTIIYP
2P2D:A|PDBID|CHAIN|SEQUENCE          TTKGHADGFDAFASPNLFPILEAGIHIHRLNTP-PAPSGEGELIVHRITPQPIGVVTIIYP
                                     : *;*****.;;:*** *******.*; . ;*:*;*.*************;;**

2OCD:A|PDBID|CHAIN|SEQUENCE          GISHEVIRNTLLQFVNAMILLTFGVGNAPQNPELLAQLMAASEHGVTVVNLIQCLAGNVN
3NTX:A|PDBID|CHAIN|SEQUENCE          GISGNVVRNFLLQFVKALIIKSYGVGNAPQNAELLDKLKKASDNGIVVVRLIQCISGSVN
EARCTC2190|PDBID|CHAIN|SEQUENCE      GISADVVRNFLAQFVKALIIRSYGVGNAPQNGEPIQVLAEASQNGIVVVNLIQCMSGRVN
2HIM:A|PDBID|CHAIN|SEQUENCE          GISADVVRNFLAQFVKALIIRSYGVGNAPQNRAFLQELQEASDNGIVVVNLTQCMSGRVN
2P2D:A|PDBID|CHAIN|SEQUENCE          GISADVVRNFLAQFVKALIIRSYGVGNAPQNRRFLQELQEASDNGIVVVNLTQCMSGRVN
                                     *** *;**.* ***;;;** ;;*******; ;; * **;**;;;********;;*.**

2OCD:A|PDBID|CHAIN|SEQUENCE          NGGVATGCALADAGVISGYDMTVEAALAKLHYLLSQNLSYEEVKKMNQQVLRGEMTL--
3NTX:A|PDBID|CHAIN|SEQUENCE          NGGYATGNALAQAGVISGFDMTVEAALTKLHYLLSQSLSPNEIRQLMQQNLRGELTDTQ
EARCTC2190|PDBID|CHAIN|SEQUENCE      NGGYATGNALAQAGVISGFDMTVEATLTKLHYLLSQNLDGAAIRNMNQQNLRGELTPDE
2HIM:A|PDBID|CHAIN|SEQUENCE          NGGYATGNALASAGVTGGADMTVEATLTKLHYLLSQELDTETIRKGMSQNLRGELTPDD
2P2D:A|PDBID|CHAIN|SEQUENCE          NGGYATGNALAHAGVTGGADMTVEATLTKLHYLLSQELDTETIRKGMSQNLRGELTPDD
                                     ********.***.****;* *** **;;**********;* .  ;; *;* *****;*
```

Fig. 4.3 MSA of type I L-asparaginases

64

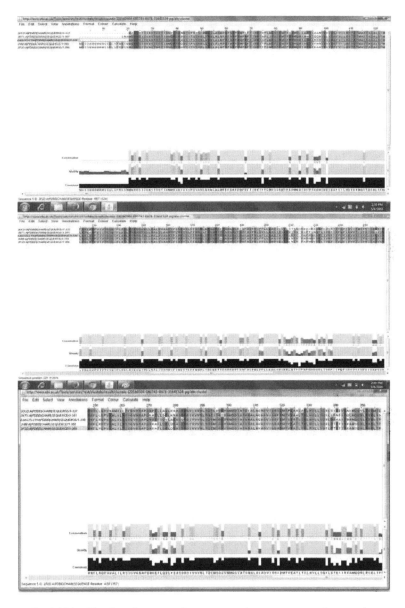

Fig. 4.4 Identification of conserved sequence regions among type I asparaginases

2OCD_A|PDBID|CHAIN|SEQUENCE 0.19401
3NTX_A|PDBID|CHAIN|SEQUENCE 0.09468
EAKCTC2190|PDBID|CHAIN|SEQUENCE 0.0348
2HIM_A|PDBID|CHAIN|SEQUENCE 0.00238
2P2D_A|PDBID|CHAIN|SEQUENCE 0.00041

Fig. 4.5 Phylogram of type I asparaginases

d. Homology modeling of Enterobacter aerogenes KCTC 2190 L-asparaginases

For predicting protein three-dimensional structure based on user provided alignment of a sequence to be modeled with recognized allied structures, homology or comparative modeling tool MODELLER 9.1 can be used. BLAST search (blast p) was performed to identify the most similar sequences whose crystal structures were experimentally determined. Search results suggested 2P2D_A, 2HIM_A and 3NTX_A as potential templates with more than 90% of sequence identity for the *Enterobacter aerogenes KCTC 2190* L-asparaginase.

Calculation of a model holding all non-hydrogen atoms was automatically done by MODELLER 9.1 using all the four templates. The tool implements comparative protein structure modeling by satisfaction of spatial restraints (Šali and Blundell, 1993), and performs various added tasks that includes de novo loop modeling, structure optimization, sequences and/or structure multiple alignment, sequence database searching, clustering, protein structure comparison etc. Best homology model was chosen based on the low discrete optimized protein energy (DOPE) score and RMSD values amongst a total of ten models predicted by MODELLER 9.1.

66

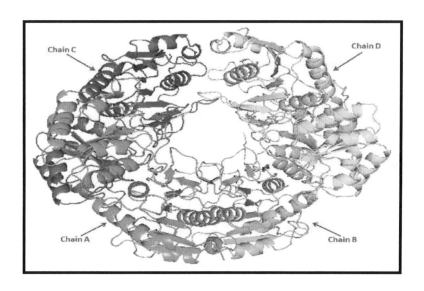

Fig.4.6 Structure of *Enterobacter aerogenes KCTC 2190* L-asparaginase developed by MODELLER 9.1

The hypothetical model of *Enterobacter aerogenes KCTC 2190/MTCC 111* L-asparaginase (Fig. 4.6) enzyme was validated using Rampage geometric evaluations to get Ramachandran plot (Fig. 4.7). The plot has 97.0% of residues present in favored, 2.7% in allowed and 0.3%residues in the outlier regions. This strongly supports the geometric fitness of the modeled enzymes. QmeanZ score of 0.09, DFire energy of -2091.25, from ERRAT 2.0 an overall quality factors of 96.851% (Fig. 4.8) indicates the good resolution modeled structure. The final validated structure was further used to study of molecular interactions with ligand molecules.

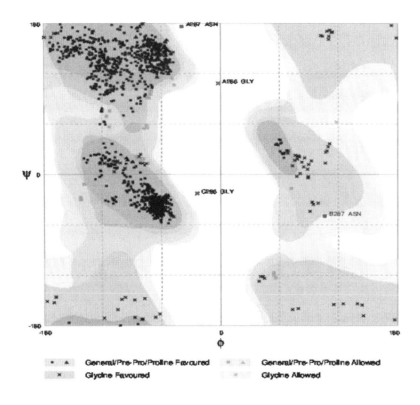

Fig. 4.7 Ramachandran plot for homology modeled *Enterobacter aerogenes KCTC 2190* L-asparaginase

68

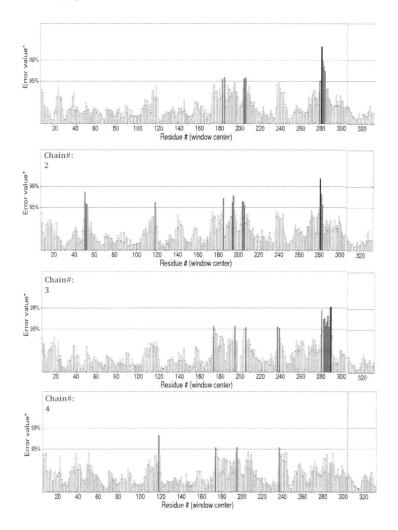

Fig. 4.8 Quality analysis for *Enterobacter aerogenes KCTC 2190* L-asparaginase by
ERRAT2.0

e. Molecular docking using Hex 8.0.0

Molecular docking was performed between the type I asparaginases and ligand molecules. All the ligands and receptors were energy minimized before docking runs using VLifeMDS. A summary of molecular docking results is described in table 4.1. Novel enzyme from *Enterobacter aerogenes KCTC 2190* has shown efficient binding with both the ligands than the other two receptor drugs with the free energies of - 172.90 KJ/mol and - 198.72 KJ/mol.

Table 4.1A summary of molecular docking results for type I L-asparaginases

S. No	Receptor	Ligand	Docking by Hex (Etotal) (KJ/mol)
1	*Enterobacter aerogenes KCTC 2190*L-asparaginase	L-Asn	-172.90
		L-Gln	-198.72
2	2HIM-Cytoplasmic L-asparaginase I from *E. coli*	L-Asn	-160.69
		L-Gln	-171.05
3	2OCD-L-asparaginase I from *Vibrio cholerae* O1 biovareltor str. N16961	L-Asn	-131.26
		L-Gln	-213.73
4	2P2D-Cytoplasmic L-asparaginase I from *E. coli*	L-Asn	-151.01
		L-Gln	-193.03
5	3NTX-L-asparaginase I from *Yersinia pestis*	L-Asn	-156.30
		L-Gln	-155.77

4.1.2 *In silico* comparison with type II enzyme drugs

a. Multiple Sequence Alignment (MSA) by CLUSTAL Omega

To produce biologically meaningful sequence alignments and to determine the evolutionary relationships, MSA was performed for *Enterobacter aerogenes KCTC 2190*L-asparaginase (EA*KCTC 2190*) and other FDA approved enzyme drugs namely L-asparaginase from *E. coli* (PDB ID: 1NNS) and L-asparaginase from *Erwinia chrysanthemi* (http://www.drugbank.ca/drugs/DB08886-Erwinaze®) (Fig. 4.9) using CLUSTAL Omega (CLUSTAL O 1.2.1). From the analysis of results, it is determined that majority of the query sequence is conserved with respective to known crystal structures (Fig. 4.10). A cladogram has been developed and it clearly says that all the sequences are evolutionarily closely related (Fig. 4.11).

70

CLUSTAL O(1.2.1) multiple sequence alignment

```
EAKCTC2190|PDBID|CHAIN|SEQUENCE    -NQKKSIYVAYTGGTIGMQN----SDHGYIPVSGHLQRQLAJPFEPHRPERPDPTIHEYA
1NNS:A|PDBID|CHAIN|SEQUENCE        ---LPNITILATGGTIAGGGDSA-TKSRYTVGKVGVENLVNAVPQLKDI--A-NNKGEQV
1NNS:B|PDBID|CHAIN|SEQUENCE        ---LPNITILATGGTIAGGGDSA-TKSRYTVGKVGVENLVNAVPQLKDI--A-NNKGEQV
Erwinase|PDBID|CHAIN|SEQUENCE      ADKLPNIVILATGGTIAGSAATGTQTTGYKAGALGVDTLINAVPEVKKL--A-NNKGEQP
                                      .*:.*****.       *   .     ::.: :*!.:         . .*

EAKCTC2190|PDBID|CHAIN|SEQUENCE    PLNDSSDNTPEDNQHIADDIKDH--YDQYDGPVILHGTDTHAFTASALSPMLENLGKPVI
1NNS:A|PDBID|CHAIN|SEQUENCE        VNIGSQDNNDNNNLTLAKKINTD--CDKTDGPVITHGTDTMEETAYPLDLTV-ICDKPVV
1NNS:B|PDBID|CHAIN|SEQUENCE        VNIGSQDNNDNNNLTLAKKINTD--CDKTDGPVITHGTDTMEETAYPLDLTV-ICDKPVV
Erwinase|PDBID|CHAIN|SEQUENCE      SNNASENMTGDAVLKLSQRVNEELARDDVDGVVITHGTDTVEESAYPLHLTV-KSDKPVV
                                     :.* .      : *    **.** .:****.    :  *****  :   :**:

EAKCTC2190|PDBID|CHAIN|SEQUENCE    VTGSQIPLAELRSDGQINLLNALYVANRYPIN--EVSLPFNNNLYHGNTTKAHADGPDA
1NNS:A|PDBID|CHAIN|SEQUENCE        NVGNPRPSTSISADGPFNLYNAVTAADKASANKGVLVNMDTVLDGRDVTKTNTTDNVAT
1NNS:B|PDBID|CHAIN|SEQUENCE        NVGNPRPSTSISADGPFNLYNAVTAADKASANKGVLVNMDTVLDGRDVTKTNTTDNVAT
Erwinase|PDBID|CHAIN|SEQUENCE      FNVNNRPNTAISADGSMNLLENVNVAGDKQSRGRGVMVVLNDKIGSARYITKTNASTLDT
                                    ..:*  *  *:**  **.*:  *:    **:*   *  :*:*:  :  *:  *

EAKCTC2190|PDBID|CHAIN|SEQUENCE    PASPNLAPLLEAG-IHIRRLGTPPAPHGHGE-LIVHPITPQ-PIGVVTIYPGISADVVRN
1NNS:A|PDBID|CHAIN|SEQUENCE        PLSNVNYGPLGYIHNGKIDYQRTPARKHTSDTPPDVSKLNELPKVGIVYNYANASDLRAKA
1NNS:B|PDBID|CHAIN|SEQUENCE        PLSNVNYGPLGYIHNGKIDYQRTPARKHTSDTPPDVSKLNELPKVGIVYNYANASDLRAKA
Erwinase|PDBID|CHAIN|SEQUENCE      PLANEEGYLGYIIGNRIYYQNRIDKLHTTRSVPDVRGLTSLPKVDILYGYQDDPEYLYDA
                                   *:: .   :*        :*      *:  *  *   :     :*    :** *

EAKCTC2190|PDBID|CHAIN|SEQUENCE    PLRQFVNVLILRSYGVGNAFQNGEPIQVLAEASQRGIAVVNLTQCMSGKVNNGGYATGNA
1NNS:A|PDBID|CHAIN|SEQUENCE        LVDAGYDGIVSA--GVGNGNLYKSVPDTLATAANTGTAVVRSSRVPTGRTTQDKEV-----
1NNS:B|PDBID|CHAIN|SEQUENCE        LVDAGYDGIVSA--GVGNGNLYKSVPDTLATAANTGTAVVRSSRVPTGRTTQDKEV-----
Erwinase|PDBID|CHAIN|SEQUENCE      AIQHGVKGIVYA--GNGAGSVSVRGIAGNRKVAPEKGVVVIRSTRTGNGIVPPDEEL-----
                                   :   . :*   .    *:*** .  :*  : :: * : *.:: :. :. .:

EAKCTC2190|PDBID|CHAIN|SEQUENCE    LAQNGVTSGPDNTVEATLTNLHYLLSQNLDGAAIRNAPQQNLRGELTPDE---------
1NNS:A|PDBID|CHAIN|SEQUENCE        --------------DDARYGPVASGTLNPQKARVLLQLALTQTKDPQQIQQIPNQY
1NNS:B|PDBID|CHAIN|SEQUENCE        --------------DDARYGPVASGTLNPQKARVLLQLALTQTKDPQQIQQIPNQY
Erwinase|PDBID|CHAIN|SEQUENCE      ---------------PGLVSDSLNPAHARILLMLALTRTSDPKVIQEYPHTY
                                      : *  *.    *  .  *  .*:. .: .  . .*
```

Fig. 4.9 MSA for *Enterobacter aerogenes KCTC 2190* L-asparaginase and other type II enzyme drugs

71

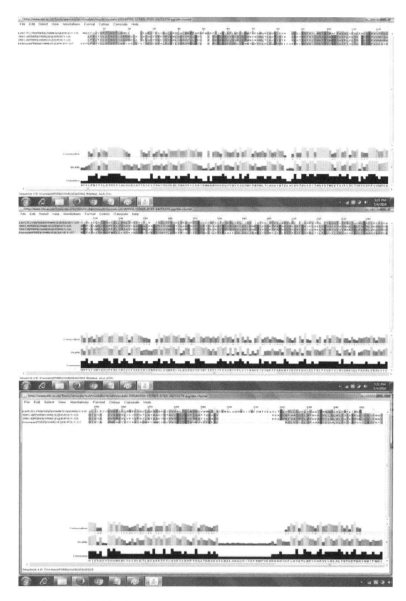

Fig. 4.10 Identification of conserved sequence regions between *Enterobacter aerogenes KCTC 2190* L-asparaginase and type II L-asparaginases

72

EAKCTC2190|PDBID|CHAIN|SEQUENCE 0.49514

1NNS_A|PDBID|CHAIN|SEQUENCE 0

1NNS_B|PDBID|CHAIN|SEQUENCE 0

Eriwinaze|PDBID|CHAIN|SEQUENCE 0.27034

Fig. 4.11 Phylogram showing evolutionary relationship between *Enterobacter aerogenes KCTC 2190* L-asparaginase and type II L-asparaginases

As the L-asparaginase from *Enterobacter aerogenes KCTC 2190* is having the conserved sequence regions with type II enzyme drugs and also having greatest binding affinity among the type I L-asparaginases, an *In silico* comparison has been performed against the type II enzyme drugs for ALL.

b. L-asparaginase from Escherichia coli (Elspar®)

As per the structural description provided by the depositors a total of nine α-helices (54-59; 85-98; 114-124; 147-159; 248-255; 273-284; 308-311; 321-331 and 338-345) and fourteen β-sheets (25-32; 69-78; 104-108; 131-134; 169-172; 175-178; 182-184; 193-195; 202-205; 208-211; 236-240; 260-265; 288-293 and 313-315) are present in it. Based on *interpro* analysis (www.ebi.ac.uk/interpro/) N-terminus of 1NNS (Fig. 4.12) has two conserved threonine residues with catalytic role.

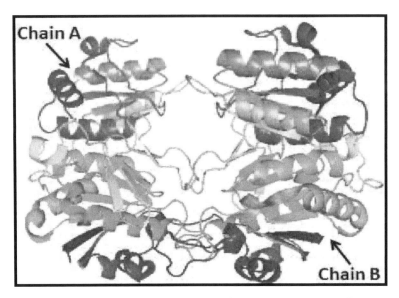

Fig. 4.12 *E. coli* L-asparaginase crystal structure (PDB ID: 1NNS)

c. Homology modeling of Erwinia chrysanthemi (Erwinaze®) L-asparaginase

BLAST search was performed against the amino acid sequence of L-asparaginase from *Erwinia chrysanthemi* (Erwinaze®) to identify the most similar sequences whose crystal structures were experimentally determined. Search results suggested 1HFJ_A, 1O7J_A, 1ZCF_A and 2JK0_A as potential templates for Erwinaze®.

The hypothetical model of Erwinaze®(Fig. 4.13) enzyme was validated using Rampage geometric evaluations to get Ramachandran plot. The plot has 97.2% and 2.8% of residues in favored and allowed regions respectively for Erwinaze® strongly supporting the geometric fitness of the modeled enzyme (Fig. 4.14). QmeanZ score of 0.671, DFire energy of -455.15, ERRAT 2.0 an overall quality factors of 84.326% for Erwinaze® L-asparaginase indicates the good resolution structure (Fig. 4.15).The final validated structure was energy minimized using the same parameters implemented in *Enterobacter aerogenes KCTC 2190* L-asparaginase minimization and the same is further used to study of molecular interactions with ligand molecule.

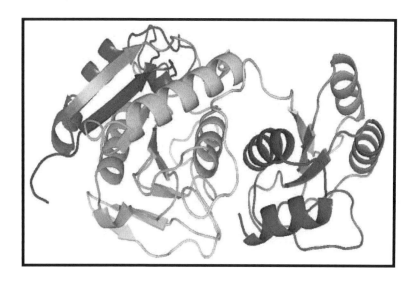

Fig.4.13 Structure of Erwinaze® developed by MODELLER 9.1

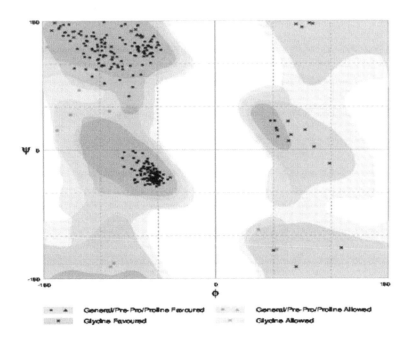

Fig. 4.14 Ramachandran plot for homology modeled *Erwinia chrysanthemi* L-asparaginase

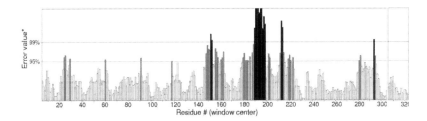

Fig. 4.15 Quality analysis for *Erwinia chrysanthemi* L-asparaginase by ERRAT2.0

d. Molecular Docking

Two ligand substrates namely L-Asn and L-Gln were docked into the catalytic site of 1NNS (Fig. 4.16). Dock runs of ligands on the enzyme were performed using HEX, iGEMDOCK and PatchDock & FireDock servers.

Docking using Hex 8.0.0

When docking runs were carried out by HEX, L-Asn and L-Gln ligand substrates resulted in the following binding energies upon the interaction with the three receptors (Table 4.2).

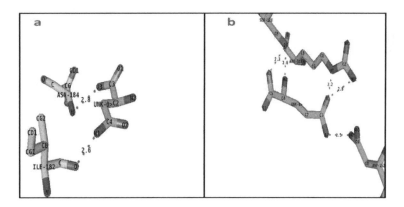

Fig. 4.16 Molecular docking of 1NNS with (a) L-asparagine (b) L-glutamine

Docking using iGEMDOCK v 2.1

Dock runs of iGEMDOCK also resulted in the similar fashion as *Enterobacter aerogenes KCTC 2190* L-asparaginase has shown very high binding affinity with L-Asn and L-Gln over the other receptors (Table 4.2) resulting in free energy values of- 60.72 kcal/mol and - 63.50 kcal/mol.

Results obtained by HEX and iGEMDOCK were validated by another protein-protein docking web server namely PatchDock followed by refinement of obtained results with FireDock server. Outcomes strongly supported the previous results with a very good binding efficiency between *Enterobacter aerogenes KCTC*

2190 L-asparaginase receptor and ligands with the calculated global energy values of -20.07 kcal/mol and -26.63 kcal/mol for L-Asn & L-Gln respectively (this value is considered to be related to free binding energy) by FireDock server.

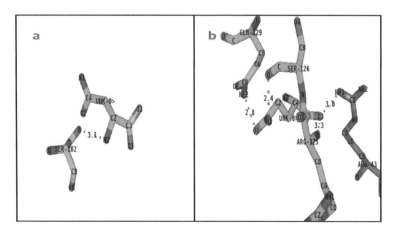

Fig. 4.17 Molecular docking of *Enterobacter aerogenes KCTC 2190* enzyme with
(a) L-asparagine (b) L-glutamine

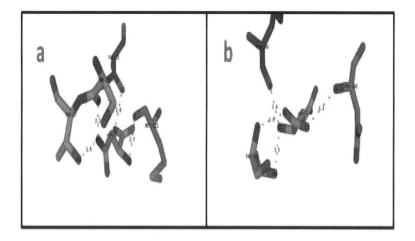

Fig. 4.18 Molecular docking of Erwinaze® with (a) L-asparagine
(b) L-glutamine

Docking using PatchDock and FireDock

Enterobacter aerogenes KCTC 2190 L-asparaginase enzyme made inter molecular hydrogen bonds with SER-102 against L-Asn ligand and on the other side generated same with ARG-43, ARG-125, SER-126 and GLN 129 against L-Gln (Fig. 4.17, Table 4.3). *E. coli* enzyme gave global energy values of -22.72 kcal/mol and -21.46 kcal/mol by forming the hydrogen bond interactions between ILE-182, ASN-184 and L-Asn; between ARG-116, SER-120 and ASP-152 and L-Gln.

Erwinaze® resulted in the hydrogen bonding pattern between THR-15, THR-26, THR-27, MET-121 and L-Asn; and between THR-165, THR-167 and ASN-180 and L-Gln (Fig. 4.18) with the binding energies of - 20.41kcal/mol and - 25.32 kcal/mol respectively. Summary of the binding energies by all the docking methods were described in table 13. The binding mode of ligands with receptor was investigated in PyMol molecular graphics viewer. All the molecular docking results clearly, strongly supported the novel *Enterobacter aerogenes KCTC 2190* L-asparaginase as a new therapeutic drug for ALL therapy over the existing ones. In order to check the stability of novel enzyme in physiological conditions molecular dynamics and simulations were performed.

Table 4.2 Summary of molecular docking results for *In silico* comparison with type II enzyme drugs

S.No	Receptor	Ligand	Docking Tool		
			Hex (Etotal) (KJ/mol)	iGEMDOCK (Kcal/mol)	PatchDock &FireDock (Kcal/mol)
1	*Enterobacter aerogenes* KCTC 2190 L-asparaginase	L-Asn	-172.90	-60.72	-23.07
		L-Gln	-198.72	-63.50	-26.63
2	*1NNS- Escherichia coli* L-asparaginase II (Elspar®)	L-Asn	-155.67	-49.51	-22.72
		L-Gln	-162.18	-57.20	-21.46
3	*Erwinia chrysanthemi* L-asparaginase (Erwinaze®)	L-Asn	-141.00	-52.64	-20.41
		L-Gln	-167.21	-57.25	-25.32

Table 4.3 Molecular interactions between L-asparaginase enzyme and ligands

S.No.	Receptor	Ligand	Residue & Atom	Ligand Atom	Bond Length (Å)
1	*Enterobacter aerogenes KCTC 2190* L-asparaginase	L-Asn	SER102 & O	N_2	3.1
			ARG43 & NH_1	O_3	3.0
		L-Gln	ARG125 & N	O_3	3.3
			SER126 & O	N_1	2.4
			GLN129 & NE_2	O_1	2.8
2	*E. coli* L-asparaginase (Elspar®)	L-Asn	ILE182 & O	N_1	2.6
			ASN184 & N	O_3	2.8
			ARG116 & NH_2	O_1	2.8
			ARG116 & NE	O_1	2.2
		L-Gln	SER120 & OG	N_2	2.6
			SER120 & OG	O_2	2.3
			ASP152 & OD_2	N_1	2.5
3	*Erwinia chrysanthemi* L-asparaginase (Erwinaze®)	L-Asn	THR15 & OG_1	O_3	2.6
			THR26 & O	N_1	2.5
			THR27 & OG_1	N_1	3.4
		L-Gln	MET121 & O	N_2	2.7
			TYR165 & O	N_2	2.6
			THR167 & OG_1	N_2	3.5
			THR167 & N	O_2	2.9
			ASN180 & N	O_3	3.3

4.1.3 Molecular Dynamics and Simulations

MD simulations were executed to check the structural behavior of apo enzyme and docked complexes in dynamic system and the trajectory analysis was done using GROMACS utilities. The total energy found from g_energy results for apo state *E. coli* (1NNS) enzyme, complex 1 and complex 2 was found to be -1.6X10^6 KJ/mol, -1.8X10^6 KJ/mol and -1.8 X10^6 KJ/mol respectively demonstrating that the models were energetically firm (Fig. 4.19).

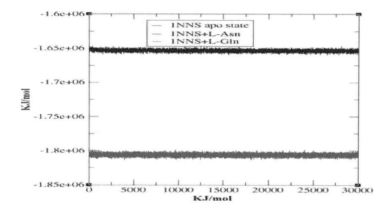

Fig. 4.19 Total energy of apo state 1NNS and its docked complexes

The structural convergence comprises terms like Root Mean Square Deviation (RMSD), Root Mean Square Fluctuation (RMSF) and Rg. The g_rms based results with backbone atoms (Fig. 4.20) exhibited early equilibration up to 5.5 ns, converged later and achieved stability with a RMSD between 0.2 nm and 0.5 nm with a peak after 26 ns for 1NNS in apo state, whereas the docked complex 1 attained equilibration at 10 ns. After the initial equilibration it was stable throughout the trajectory oscillating between RMSD values of 0.4 nm and 0.5 nm which were a little higher than apo state enzyme. Though complex 2 initially followed apo state enzyme, achieved a stable conformation throughout the simulation soon after the first 4 ns with an average backbone RMSD between 0.26 nm to 0.42 nm. By showing much higher consistency throughout the entire trajectory docked complex 2 confirmed its stable conformation over both the apo state enzyme and complex 1.

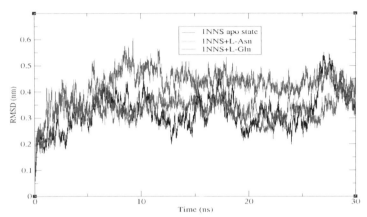

Fig. 4.20 RMSD of backbone atoms for apo state 1NNS and its docked
complexes

The examination of g_rmsf results presented the oscillations with Cα atoms
with respect to residues. Results of g-rmsf analysis revealed few fluxes in Cα atoms
when compared to apo enzyme residues. As per RMSF plot, residues of complex 1
are stable (Fig. 4.21a) with few peaks of more than 0.25 nm compared to complex 2
(Fig. 4.21b) where many residues are oscillating around 0.3 nm along with one high
peak at the last one with 0.4 nm. The comparison of RMSF outcomes showed minor
variations in ligand binding sites (Table 4.4) and their effect on complex formation.

Table 4.4 Comparison of RMSF values for 1NNS and its substrates

S.No.	Residue	RMSF (nm)		
		1NNS in apo state	Complex 1	Complex 2
1	ARG116	0.0875	---	0.1016
2	SER120	0.0902	---	0.1305
3	ASP152	0.1667	---	0.1568
4	ILE182	0.1372	0.1295	---
5	ASN184	0.1684	0.2356	---

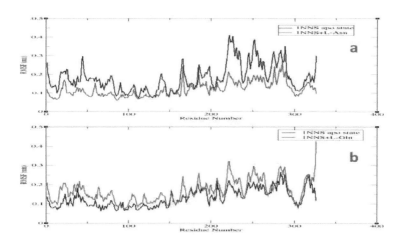

Fig. 4.21 RMSF of Cα atoms for apo state 1NNS and its docked complexes

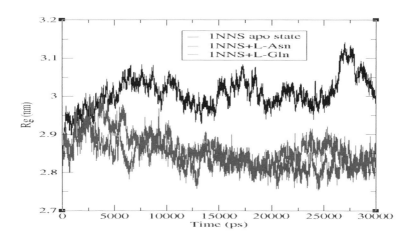

Fig. 4.22 Radius of gyration plot of Cα atoms for apo state 1NNS and its docked complexes

Radius of gyration (Rg) describes overall spread of molecule and was calculated using g_gyrate tool of GROMACS. Continuous drifts in gyration radii for apo state enzyme in the range of nearly 2.9 nm and 3.15 nm describes the dynamic nature of 1NNS. Docked complexes are better in terms of conformational stability with constant and low Rg values authorizing their better folding (Fig. 4.22) than unbound enzyme. Complex 1 was stable during entire MD run after 13 ns with no further drifts till end, and the second complex was stable only after 6 ns. The H-bond analysis of docked complexes revealed the pattern of hydrogen bonding during the course of 30 ns simulations (Fig. 4.23). Inter-hydrogen bond linking configuration was evidently elucidated by the binding of L-Asn (Fig. 4.23a) and L-Gln (Fig. 4.23b) in the apo enzyme initially with large number of hydrogen bonds which were reduced to 2 to 3 after conformational change due to binding after 12 ns simulations.

Though MD simulation outcomes were well correlated with outcomes of molecular docking with two to five h-bonds observed in dynamic state, surprisingly loss of hydrogen bonds at the end of simulations for complex 1 was observed. Overall, h-bond interactions revealed the kinship of ligand molecules with *E. coli* L-asparaginase and its dual activity. An overall result from this computational study confirms the dual functionality of the *E. coli* L-asparaginase enzyme. The instability of the enzyme upon the interaction with the ligand substrates based on RMSD, RMSF and H-bond analysis suggests the need for identifying a new and stable L-asparaginase enzyme from diverse source with similar therapeutic effects for better curing of ALL.

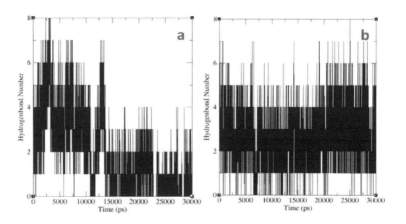

Fig. 4.23 H-bond plot for (a) 1NNS+L-Asn (b) 1NS+L-Gln

84

Similarly MD simulations were performed for modeled Erwinaze® enzyme and docked complexes in the dynamic system. The total energy of - 9.3X10⁵ KJ/mol for Erwinaze® modeled enzyme demonstrates the model's stability (Fig. 4.24). In the g_rms based results with backbone atoms (Fig. 4.25) apo state enzyme is stable right from the launch of MD run with a drift at 10 ns followed by a stable confirmation till the finale of entire run.

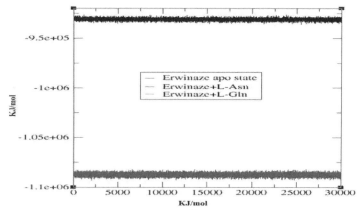

Fig. 4.24 Total Energy of Erwinaze® apo enzyme and its docked complexes

In case of docked complexes, though the complex 3 is unwavering till 20 ns it exhibited an increasing path later and extended the RMSD value to virtually 0.4 nm. RMSD of complex 3 with L-Asn till 10 ns specified the ligand binding in apo state and later a significant conformational change with an increasing value was observed which might be the plausible cause to obtain L-Asn bound state. Its high RMSD value shows the instability of the complex which is undesirous. On the other hand complex 4 has shown a perpetual RMSD value around 0.3 nm showing its high stability in the 300°K atmosphere supporting the experimental results of neurotoxicity of L-glutaminase nature of enzyme. This leaves the plasma pool L-Asn free for the uptake by cancerous cells on one hand because of unstable nature of complex 3, and causing the toxicity in normal cells due to its stable glutaminase activity. Both the complexes are displaying their undesirable results against the actual desired L-asparaginase role of enzyme in cancer therapy.

85

In the RMSF result analysis many oscillations in the residues with Cα atoms were observed in case of complex 3 with a fall and amendment in initial peak due to the binding of L-Asn in that region. This substrate binding also influenced the entire enzyme with several numbers of residual fluctuations with a near RMSF value of 0.3 nm (Fig. 4.26). The same plot also disclosed the presence of a second high peak in the vicinity of bounded residues establishing H-bonds with ligand leading to a very high RMSF value of around 0.6 nm in complex 4, leaving most of the other residues stable. The comparison of RMSF outcomes exhibited minor variations in ligand binding sites (Table 4.5) and their effect on complex formation.

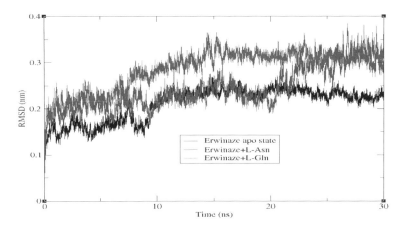

Fig. 4.25 RMSD plot for apo state Erwinaze® enzyme and its docked complexes

Table 4.5 Comparison of RMSF values for Erwinaze® and its substrates

S.No.	Residue	RMSF (nm)		
		Erwinaze® in apo state	Complex 3	Complex 4
1	TYR15	0.0905	0.1108	---
2	THR 26	0.5626	0.4778	---
3	THR27	0.4196	0.4855	---
4	MET121	0.1318	0.1355	---
5	TYR165	0.0647	---	0.0677
6	THR167	0.0616	---	0.0722
7	ASN180	0.1866	---	0.1255

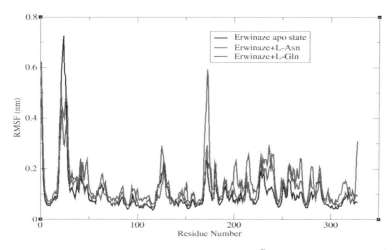

Fig. 4.26 RMSF plot for apo state Erwinaze® enzyme and its docked complexes

Radius of gyration (Rg) was done to analyze the structural compactness of enzyme and was calculated using g_gyrate tool of GROMACS. Though the modeled enzyme in its free state has some structural compactness, bounded complexes were shown their fluctuating nature from Rg plots. Complex 3 has displayed the drifts during entire MD run in the array of 2.0 nm and 2.08nm. In the plot of complex 4 it revealed its compactness with a little drift at 15 ns time point with the Rg values oscillating between 2.0 nm and 2.06 nm (Fig. 4.27). In order to authenticate the hydrogen bonding pattern docking results, H-bond analysis was also executed along with the intact MD trajectories (Fig. 4.28). Surprisingly the desired stable H-bonding pattern for complex 3 described missing linkage of hydrogen bonds between receptor and ligand. During 7.5 ns point the H-bonds were increased to 5, but after 4.03 ns of trajectory H-bonds were missed and again formed from 6 ns.

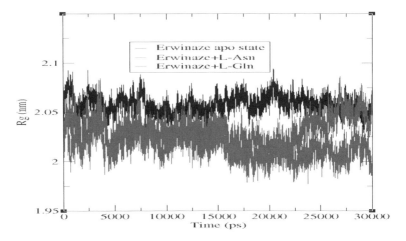

Fig. 4.27 Rg plots for apo state Erwinaze® enzyme and its docked complexes

The same was again repeated at around 9 ns and 11 ns time points. Finally after 12.6 ns there were no h-bonds, showing the disappearance of H-bond linkage between enzyme and substrate (Fig. 4.28a). The loss of intermolecular hydrogen bonds may induce a spatial conformational change in the tertiary structure of the enzyme i.e., unstable structure of docked complex. This becomes a major shortcoming for the enzyme in therapeutic perspective, supporting possibility of its lower effect on cancer cells. On the other hand it has shown the constant H-bond pattern with L-Gln throughout MD run with an average of 7 H-bonds throughout entire trajectory (Fig. 4.28b).

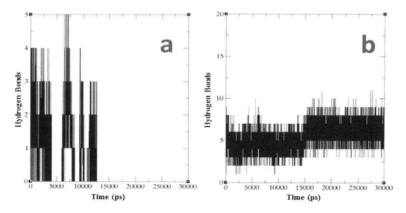

Fig. 4.28 H-bond plot for (a) Erwinaze®+L-Asn (b) Erwinaze®+L-Gln

H-bond pattern of both complexes did not draw a parallel with the docking result and this gives scope for the further investigation for efficient and firm drug in ALL therapy. Entire computational analysis describes the necessity of further intensive investigation on other sources of L-asparaginase enzyme which is more effective on cancerous cells with fewer side effects.

Molecular dynamics and simulations were also performed for *Enterobacter aerogenes KCTC 2190* L-asparaginase in apo state along with its docked complexes and the trajectory was analyzed. The total energy for apo state enzyme, complex 5 and complex 6 was found to be -2.3X106 KJ/mol, -2.4X106 KJ/mol and -2.4X106 KJ/mol respectively signifying stability of all the molecules (Fig. 4.29). The g_rms based results with backbone atoms (Fig. 4.30) exhibited initial equilibration up to 7.5 ns, underway to converge later and achieved stability with a constant RMSD value of around 0.16 nm throughout the entire 30 ns simulation in apo state, whereas the docked complex 5 attained an early equilibration at 6 ns. After the initial equilibration it was stable throughout the trajectory oscillating between RMSD values of0.25 nm and 0.30 nm which were higher than apo state enzyme.

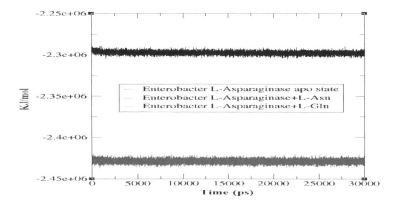

Fig. 4.29 Total energy of apo state *Enterobacter aerogenes KCTC 2190*
L-asparaginase and its docked complexes

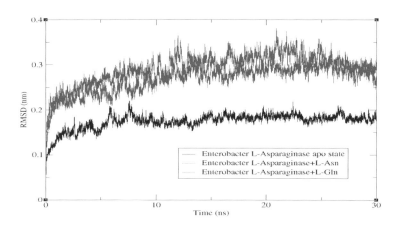

Fig. 4.30 RMSD plot of apo state *Enterobacter aerogenes KCTC 2190*
L-asparaginase and its docked complexes

But the above RMSD values were permissible for this docked complex as per
the GROMACS tutorial. But the RMSD pattern of complex 6 was quite different
from the above two, achieved stability only after 11.5 ns point and highly oscillating
till the end of trajectory with an average backbone RMSD between 0.2 nm to 0.36
nm. By showing much higher consistency throughout the entire trajectory docked

complex 5 confirmed its stable conformation over the other complex. The examination RMSF results presented the oscillations with Cα atoms with respect to residues. Results revealed few fluxes in Cα atoms of docked complexes with respect to apo enzyme residues. As per RMSF plot residues of complex 5 were stable with a high peak in the beginning with an RMSF of 0.45 nm and few others around 0.3 nm which might be due to the binding of L-Asn with SER102 of apo enzyme. In the similar fashion few peaks were observed for complex 6 (Fig. 4.31) where many residues were stable and with few peaks in the close vicinity of bounded residues oscillating around 0.35 nm.

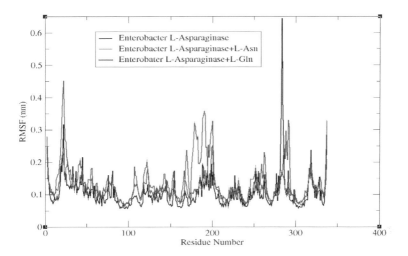

Fig. 4.31 RMSF plot of apo state *Enterobacter aerogenes KCTC 2190* L-asparaginase and its docked complexes

The comparison of RMSF outcomes showed minor variations in ligand binding sites (Table 4.6) and their effect on complex formation. Radius of gyration (Rg) describes overall spread of molecule and was calculated using g_gyrate tool of GROMACS. Continuous drifts in gyration radii for apo state enzyme till 17.5 ns in the range of nearly 3.4 nm and 3.47 nm describes the dynamic nature of apo enzyme of *Enterobacter aerogenes KCTC 2190*. Thereafter it was stable till the end. Docked complexes were better in terms of conformational stability with low Rg values authorizing their better folding (Fig. 4.32) than unbound enzyme.

Table 4.6 Evaluation of RMSF values for *Enterobacter aerogenes KCTC 2190*
L-asparaginase molecules

S.No.	Residue	RMSF (nm)		
		***Enterobacter* L-asparaginase in apo state**	**Complex 5**	**Complex 6**
1	ARG43	0.1301	---	0.1823
2	SER102	0.0656	0.0828	---
3	ARG125	0.0974	---	0.1128
4	SER126	0.0914	---	0.1042
5	GLN129	0.0905	---	0.0946

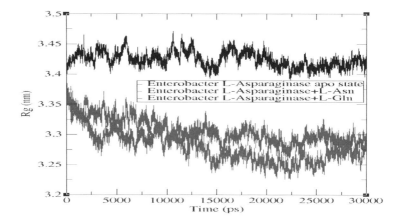

Fig. 4.32 Rg plot of apo state *Enterobacter aerogenes KCTC 2190* L-asparaginase
and its docked complexes

Complex 5 was stable during entire MD run after 2.6 ns with few small drifts
till end ranging around 3.3 nm, and the second complex was stable only after 15 ns
with an average gyration value of 3.26 nm thereafter. The h-bonding interactions
between the docked complexes established the pattern of h-bonding during entire 30
ns of trajectory (Fig. 4.33). Binding of L-Asn (Fig. 4.33a) and L-Gln (Fig. 4.33b)
with the apo state enzyme clearly explained interaction pattern of inter-hydrogen
bonding.

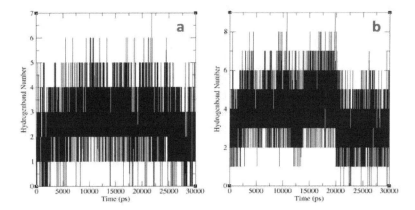

Fig. 4.33 H-bond plot for *Enterobacter aerogenes KCTC 2190* L-asparaginase
docked complexes with   (a) L-Asn   (b) L-Gln

Molecular docking outcomes were well correlated with MD simulation results with an average of 2 to 4 and 3 to 5 hydrogen bonds for complex 5 and complex 6 respectively. Overall results revealed the affinity of ligand molecules with novel L-asparaginase. Overall result of this computational study confirms the stability of desired L-asparaginase + L-Asn complex over the L-glutamine complex. The stability of the enzyme upon the interaction with the L-Asn ligand substrate over the L-glutamine complex based on computational results including RMSD, RMSF and H-bond analysis strongly supports the novel *Enterobacter aerogenes KCTC 2190* L-asparaginase as a new potent therapeutic molecule with similar therapeutic effects for better curing of ALL.

4.2 *In vitro* studies on L-asparaginase

4.2.1 Qualitative assay for L-asparaginase production

Qualitative analysis for the L-asparaginase enzyme production was done by designing a system consisting of two control plates and one test plate having L-asparagine and Phenol Red indicator in medium. Control plate 1 was uninoculated and control plate 2 and test plates are not. Absence of L-asparagine in control plate 2 will not lead to any color conversion. Development of pink color in test plate confirms L-asparaginase production by the bacterium (Fig. 4.34). The same was reconfirmed following the protocol by Gulati et al., (Gulati, et al., 1997).

Control 1	Control 2	Test
Without Organism	Without L-asparagine	With Organism and L-asparagine

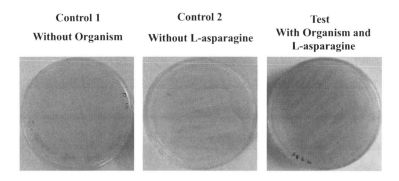

Fig. 4.34 Rapid plate assay for Qualitative analysis of L-asparaginase production

4.2.2 Optimization of L-asparaginase by OFAT method

a. Influence of batch time

Batch time on synthesis of L-asparaginase was evaluated by using production medium in a shake flask incubated at an agitation rate of 200 rpm and at 30°C temperature. After every 6 h, 10 ml culture was introverted and L-asparaginase activity profile was measured using crude sample. Results showed that 30 h as optimum time of incubation for enzyme production (2.92 IU ml^{-1}). However, further incubation led to the minimized yield of L-asparaginase activity. Total protein was peak at 18 h (373 µg/ml) which was reduced to 191.7 µg/ml at 30 h of time point. Microbial growth pattern was recorded throughout the incubation period (Fig. 4.35).

94

The novel organism gave higher levels of L-asparaginase activity compared to *Enterobacter cloacae* [1.3 IUml^{-1}] (Sharma and Husain, 2015), *mollusks* [0.75 IUml^{-1}] (Benny, 2010) and *Streptomyces sp.* [1.56 IUml^{-1}] (Selvam and Vishnupriya, 2013).

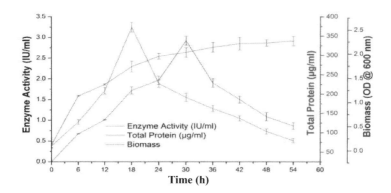

Fig. 4.35 Influence of incubation time on L-asparaginase production

b. Influence of pH and Temperature

The enzyme shows the greatest activity of 2.9 IUml^{-1} at pH 8. In lesser and higher pH conditions enzyme activity was stumpy compared to pH 8. In contrast to enzyme activity, total protein content was maximum (667.8 µg/ml) at 18 h in pH 7 culture. pH and growth profiles were recorded throughout the time course (Fig. 4.36). Even though Sharma and Husain et al., obtained an activity value of 3.7 IUml^{-1} at pH 7, it was decreased to 2.45 IUml^{-1} at pH 8 (Sharma and Husain, 2015). The activity levels are almost equivalent to the results obtained by using *Streptomyces* cultures (Selvam and Vishnupriya, 2013).

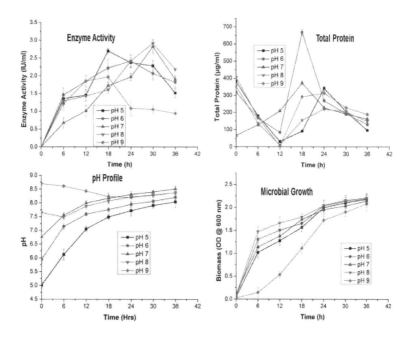

Fig. 4.36 Effect of pH on L-asparaginase production

Using this optimum pH (8.0), influence of temperature on L-asparaginase activity was studied by maintaining the inoculated cultures at different temperatures (20°C, 25°C, 30°C and 35°C). Enzyme activity was doubled (5.8 IUml⁻¹) for the sample incubated at 25°C at 30 h of time. Maximum total protein of 216.8µg/ml was achieved for the same sample at 30 h incubation compared to other samples at different temperatures. pH and growth profiles were also shown in figure 4.37. The experimentally determined enzyme activity of 2.9 IUml⁻¹ was quite higher than the activity values obtained by other researchers using different fungi [5.38 IUml⁻¹, 3.06 IUml⁻¹] (Hosamani and Kaliwal, 2011; Selvam and Vishnupriya, 2013), bacteria [4.8 IUml⁻¹, 3.4 IUml⁻¹ and 3.7 IUml⁻¹] (Kumar, et al., 2012; Mahajan, et al., 2012; Sharma and Husain, 2015) and mollusks [2.7 IUml⁻¹] (Benny, 2010).

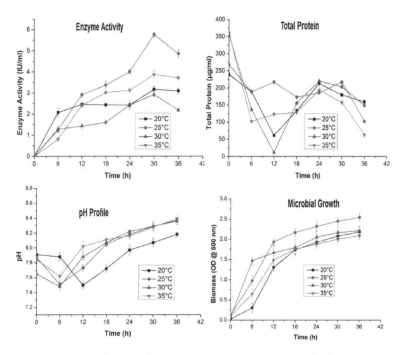

Fig. 4.37 Effect of temperature on L-asparaginase production

c. Influence of carbon sources on L-asparaginase production

Trisodium citrate, glucose, starch and sucrose were used at 0.75 % as sole carbon sources against *Enterobacter aerogenes MTCC 111* for L-asparaginase production. Among the four, trisodium citrate gave maximum activity of 5.8 IUml[-1] which was same to previous experiment and higher than activity obtained by Mukherjee [0.7 IUml[-1]] (Mukherjee, et al., 2000). Whereas only 1.85 IUml[-1] of L-asparaginase activity was obtained using 1% of Sucrose as carbon source (Kumar, et al., 2012). In contrast, Glucose and Pyruvate reported 29.86 IUml[-1] and 7.01 IUml[-1]of activities with *Aspergillus sp.* (Gurunathan and Sahadevan, 2011)and *Enterobacter cloacae*(Sharma and Husain, 2015). But Glucose gave a remarkable total protein content of 1012.6 µg/ml at 24 h of incubation (Fig. 4.38). Fig. 4.38 describes the pattern of pH variation and growth profile.

Though the microbial growth was high in medium where glucose as carbon source, the L-asparaginase activity was much more dependent on pH and trisodium citrate.

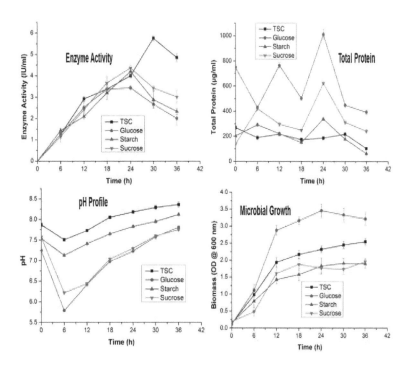

Fig. 4.38 Effect of different carbon sources on L-asparaginase production

Throughout the incubation period very high total protein quantities were observed over the other sources. Then concentration of trisodium citrate was optimized by inoculating bacterium in production media with 0.75, 1.5 and 2.0 % of carbon sources. But there was no improvement in enzyme activity and total protein with the increased concentration of trisodium citrate during the time course. Recorded profiles of pH and bacterial growth were also shown in figure 4.39.

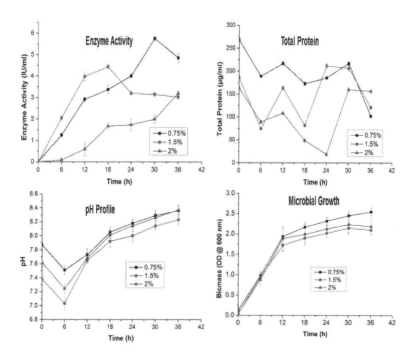

Fig. 4.39 Effect of different concentrations of Trisodium citrate on L-asparaginase production

d. Influence of L-asparagine on L-asparaginase production

Though the different concentrations of L-asparagine concentrations were tested highest enzyme activity of enzyme was observed in culture with 1% of inducer. 5.8 IUml[-1] of enzyme activity was observed at 30 h incubation with experimentally derived optimum conditions of pH 8, 25°C temperature and 0.75 % of carbon source. On the other side total protein content was comparatively low for 1% culture. pH and growth profile were also shown in figure 4.40.The observed enzyme activity at 1% L-Asparagine concentration is comparatively good compared with other bacterial sources (Kumar, et al., 2012; Mahajan, et al., 2012). It was comparably equal to *Aspergillus* culture yields (Gurunathan and Sahadevan, 2011). Though *Bacillus licheniformis* gave 8.4 IUml[-1] at 18 h, (Mahajan, et al., 2012) it consumed 30% of L-asparagine in media which is an economical drawback from an industrial point of view.

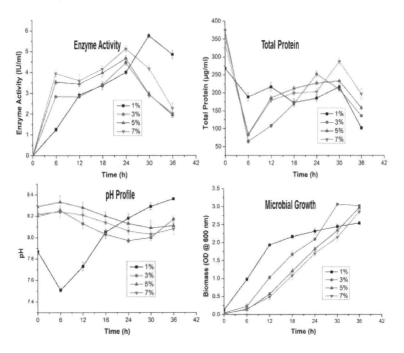

Fig. 4.40 Effect of L-asparagine on L-asparaginase production

e. Influence of nitrogen sources on L-asparaginase production

Ammonium chloride (0.15%) was proved to be the best nitrogen source over the others (yeast extract, beef extract and casein) tested against *Enterobacter aerogenes MTCC 111.* Activity of L-asparaginase increased to 6.43 IUml^{-1} at 24 h time point in the preliminary study with low content of total protein in the broth throughout the time course. Recorded values of pH and microbial growth were given in figure 4.41.

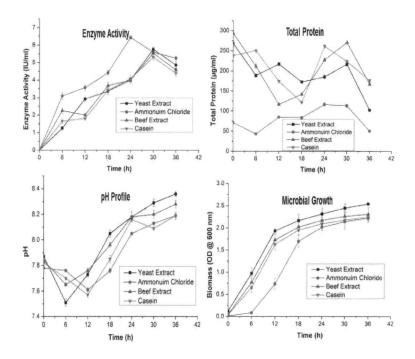

Fig. 4.41 Effect of different nitrogen sources on L-asparaginase production

To define optimum concentration of Ammonium chloride for attaining increased enzyme activity production media was supplemented with 0.15 %, 0.30 %, 0.45 % and 0.60 %. Optimum parameters were maintained based on the previous runs. Culture with 0.6% of nitrogen source gave a remarkable enzyme activity value of 10.6 IUml^{-1} at 30 h of incubation time. Reduced time of enzyme production with highest activity is an important factor considered in commercial production of enzyme. Total protein, pH and growth patterns were also recorded (Fig. 4.42). On the other hand, Yeast extract gave only 0.75 IUml^{-1} and 1.39 IUml^{-1} with *mollusks* (Benny, 2010) and *Streptomyces sp.* (Selvam and Vishnupriya, 2013). A combination of Tryptone 1% with Yeast extract 0.5% resulted in only 1.92 IUml^{-1} of activity value with *E.coli K12* (Kumar, et al., 2012). The novel organism gave even higher values compared to *Enterobacter cloacae* cultures (Sharma and Husain, 2015).

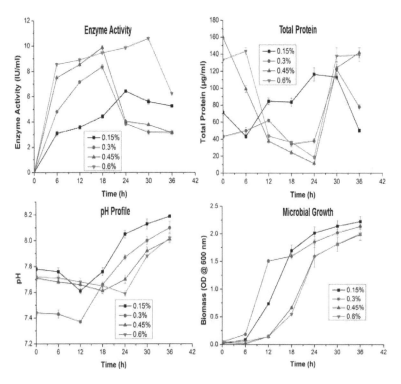

Fig. 4.42 Effect of different concentrations of Ammonium chloride on
L-asparaginase production

f. Effect of inoculum size on L-asparaginase production

Finally different sizes of inoculum (0.5%, 1.0%, 1.5% and 2.0%) were added
to 250 ml culture flasks containing 50 ml of production media with optimum carbon,
nitrogen and inducer concentrations based on the earlier results. Increase in size of
inoculum reduced the enzyme activity. Among the tested inoculum size 0.5% was
proved as best with 10.6 IUml[-1] activity value by standard L-asparaginase activity
assay at 30 h of time point. Only culture with 1.5 % of inoculum gave the second
maxima of 8.79 IUml[-1] at 30 h of time course. Increased inoculum sizes gave higher
quantities of total protein values compared to 0.5% inoculum size (Fig. 4.43). pH and
growth profiles were measured for every 6 h of incubation time. This novel microbe
reported better enzyme activity values at lower inoculum sizes compared to a value

of 1.02 IUml[-1] with the use of 10% inoculum size by Kumar, et al., 2012 and on the other hand *Enterobacter aerogenes KCTC 2910* gave more than 5 IUml[-1] which was quite higher than an experimental value of 2.46 IUml[-1] at 2% inoculum size obtained by Sharma and Husain, 2015.

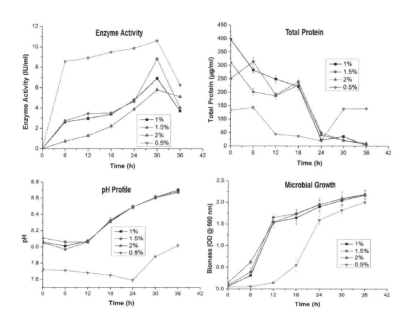

Fig. 4.43 Effect of different inoculum sizes on L-asparaginase production

4.2.3 Statistical Optimization of L-asparaginase production

a. Plackett-Burman design for screening of process parameters

In the Plackett–Burman design a sum of fourteen process parameters were examined with respect to their influence on production of L-asparaginase (Table 3.5). Table 4.7 describes the design matrix chosen for screening of major parameters for enzyme production and the resultant responses. Then calculation of model adequacy was done, and the process variables showing substantial effects were screened based on Student's t-test (Table 4.8). Factors having P-values lesser than 0.05 were considered to have momentous influence on response, and were therefore considered for next phase of optimization. Temperature was identified to be the greatest significant component with P-value 0.001, trailed by pH (0.003), L-asparagine (0.019), agitation (0.021), Ammonium chloride (0.028) and Tri Sodium Citrate (0.046). The poorer is the probability value higher is its significance on the production of L-asparaginase. Amongst the three screened significant physical variables, only agitation had a positive effect, whereas on the other side among the significant chemical variables, Tri Sodium Citrate and L-asparagine exerted positive effects on L-asparaginase production. Among all the significant variables identified, only the physical variables (Temperature, pH and rpm) were considered for further optimization by an RSM design. Further all the media components were optimized and the optimized values were given in table 4.9. A Pareto chart is drawn describing the standardized effects of process variables for L-asparaginase activity (Fig. 4.44).

Table 4.7 Plackett-Burman design table used for screening of variables

S.No	Time (h)	Temp (°C)	pH	rpm	DO	Inn Size (ml)	Inn Age (h)	TSC (%)	DAHP (%)	MS (%)	FS (%)	DPHP (%)	AC (%)	L-ASN (%)	L-asparaginase Activity (IU ml⁻¹)
1	+1	+1	+1	+1	+1	-1	-1	-1	-1	-1	-1	+1	-1	+1	0.71±0.002
2	-1	+1	+1	-1	+1	+1	+1	-1	-1	+1	-1	-1	+1	+1	1.7±0.0010
3	+1	+1	-1	+1	+1	+1	-1	+1	-1	-1	+1	-1	+1	+1	5.92±0.003
4	-1	+1	+1	+1	+1	-1	-1	+1	+1	-1	+1	-1	-1	-1	46±0.001
5	-1	-1	+1	+1	+1	-1	+1	+1	+1	-1	-1	+1	+1	+1	14.48±0.015
6	-1	+1	-1	+1	+1	-1	-1	+1	+1	+1	-1	-1	+1	+1	5.61±0.008
7	+1	+1	-1	-1	-1	-1	+1	+1	-1	+1	+1	+1	+1	+1	2.00±0.004
8	-1	-1	+1	-1	-1	+1	-1	+1	+1	+1	+1	-1	-1	-1	5.81±0.009
9	+1	+1	-1	+1	-1	+1	+1	-1	+1	+1	+1	-1	-1	-1	4.28±0.013
10	-1	-1	+1	+1	-1	+1	+1	-1	-1	+1	+1	+1	-1	+1	8.60±0.002
11	+1	+1	+1	+1	-1	+1	+1	+1	-1	-1	+1	-1	-1	+1	2.01±0.008
12	+1	-1	-1	-1	+1	+1	-1	+1	+1	+1	+1	+1	-1	-1	5.35±0.001
13	+1	+1	+1	-1	-1	-1	-1	+1	-1	+1	+1	+1	+1	+1	5.18±0.005
14	+1	-1	+1	+1	+1	-1	+1	-1	+1	-1	+1	-1	+1	+1	2.50±0.004
15	-1	-1	+1	+1	+1	-1	+1	-1	+1	+1	-1	-1	-1	-1	9.00±0.006
16	+1	-1	+1	+1	+1	+1	-1	-1	+1	+1	-1	+1	+1	-1	2.00±0.011
17	-1	-1	-1	+1	+1	-1	-1	-1	-1	-1	+1	-1	-1	-1	7.92±0.005
18	+1	-1	-1	-1	+1	+1	+1	-1	-1	-1	-1	+1	-1	+1	15.71±0.012
19	-1	+1	-1	+1	+1	+1	+1	-1	-1	+1	+1	-1	+1	-1	2.99±0.016
20	-1	+1	+1	-1	-1	+1	+1	+1	+1	-1	-1	+1	+1	-1	1.02±0.005

Table 4.8 Estimated effects and coefficients for L-asparaginase activity

Term	Effect	Coef	SE Coef	T Value	P Value
Constant		5.370	0.3532	15.20	0.000
Time	-1.605	-0.802	0.3532	-2.27	0.072
Temp	-4.572	-2.286	0.3532	-6.47	0.001[a]
pH	-3.912	-1.956	0.3532	-5.54	0.003[a]
rpm	2.352	1.176	0.3532	3.33	0.021[a]
DO	1.182	0.591	0.3532	1.67	0.155
Inn Size	-0.063	-0.032	0.3532	-0.09	0.932
Inn Age	1.522	0.761	0.3532	2.15	0.084
TSC	1.862	0.931	0.3532	2.64	0.046[a]
DAHP	0.119	0.059	0.3532	0.17	0.873
MS	-0.547	-0.274	0.3532	-0.77	0.473
FS	0.132	0.066	0.3532	0.19	0.859
DPHP	0.182	0.091	0.3532	0.26	0.807
AC	-2.173	-1.087	0.3532	-3.08	0.028[a]
L-ASN	2.406	1.203	0.3532	3.41	0.019[a]
[a] codes for significant variable					

Table 4.9 Optimized media components by Plackett-Burman screening

S.No	Variable	Concentration (%)
1	TSC	1
2	DAHP	1
3	MS	0.01
4	FS	0.005
5	DPHP	0.05
6	AC	0.5
7	L-ASN	2

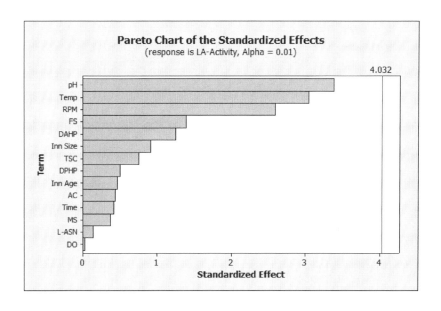

Fig. 4.44 Pareto chart of the standardized effects of process variables for
L-asparaginase activity

b. Model development and optimization of bioprocess parameters by FCCCD

The employment of RSM generated the subsequent quadratic regression
equations for two objectives i.e., L-asparaginase and L-glutaminase enzyme
activities [Eq. (3) and Eq. (4)]. Table 10 denotes the variety of process parameters;
the design of experiments and the outcome achieved for both the objectives were
given in table 4.10. The information obtained in this study described that the ultimate
enzyme activity was contingent on the combination of Temperature, rpm and pH.
The second order polynomial equations uttered as coded values were fitted to the
investigational data of the FCCCD for forecasting enzyme activities as given in Eq.
(3) and Eq. (4). Experimental outcomes coupled to the processing set of every
autonomous parameter are described in table 4.10 and coefficients of regression
equation were calculated. Evaluation of the predicted values with the experimental
results indicated that both the data were in sensible concurrence. Optimized values to

enhance final responses were identified as 25, 100 and 7.96 for Temperature, rpm and pH respectively.

Modeling equations described in terms of coded factors were as follows

$$Y_1 = +20.21 - 3.41*A + 1.34*B - 4.31*C - 0.43*A*B + 2.34*A*C - 1.64*B*C - 1.96*A^2$$
$$-1.15*B^2 - 4.78*C^2 \qquad\qquad\longrightarrow \quad (3)$$

$$Y_2 = +4.60 - 0.59*A + 0.19*B - 0.90*C - 0.88*A*B + 0.48*A*C + 0.058*B*C - 1.22*A^2$$
$$-0.59*B^2 + 0.24*C^2 \qquad\qquad\longrightarrow \quad (4)$$

The outcomes of present study illustrated that the final responses were reliant on the blend of Temperature, rpm and pH. The second-order polynomial equation fitted to the investigational data of CCD (described as coded values) for enzyme activity predictions were given in Eq. (3) and Eq. (4) where Y_1 is L-asparaginase activity and Y_2 is L-glutaminase activity.

Table 4.10 RSM design table for optimization of L-asparaginase and L-glutaminase activities with observed and predicted values

Std Order	Run Number	Temp (°C)	rpm	pH	L-asparaginase Activity (IU/ml⁻¹)			L-glutaminase Activity (IU/ml⁻¹)		
					Observed	Predicted	Residual	Observed	Predicted	Residual
1	17	-1	-1	-1	19.15±0.003	18.96	0.19	3.79±0.001	3.97	-0.18
2	18	+1	-1	-1	7.57±0.002	8.34	-0.77	3.59±0.001	3.61	-0.02
3	19	-1	+1	-1	24.61±0.005	25.79	-1.18	5.98±0.003	5.99	-0.01
4	11	+1	+1	-1	13.98±0.001	13.45	0.53	1.89±0.002	2.11	-0.22
5	16	-1	-1	+1	8.48±0.004	8.95	-0.47	1.25±0.003	1.12	0.13
6	7	+1	-1	+1	8.91±0.003	7.67	1.24	2.57±0.001	2.65	-0.08
7	10	-1	+1	+1	10.05±0.025	9.21	0.84	3.29±0.002	3.36	-0.07
8	15	+1	+1	+1	6.09±0.001	6.21	-0.12	1.48±0.001	1.38	0.10
9	3	-1	0	0	22.27±0.026	21.65	0.62	4.09±0.001	3.96	0.13
10	4	+1	0	0	13.96±0.021	14.84	-0.88	3.01±0.001	2.79	0.22
11	14	0	-1	0	17.53±0.015	17.72	-0.19	3.97±0.002	3.82	0.15
12	9	0	+1	0	20.32±0.012	20.40	-0.08	4.39±0.011	4.19	0.20
13	1	0	0	-1	20.97±0.017	19.74	1.23	6.15±0.013	5.73	0.42
14	5	0	0	+1	9.61±0.008	11.11	-1.50	3.86±0.002	3.94	-0.08
15	2	0	0	0	20.45±0.006	20.21	0.24	4.81±0.003	4.60	0.21
16	6	0	0	0	20.93±0.014	20.21	0.72	4.35±0.001	4.60	-0.25
17	13	0	0	0	20.19±0.013	20.21	-0.02	3.99±0.002	4.60	-0.61
18	20	0	0	0	20.42±0.010	20.21	0.21	4.61±0.004	4.60	0.01
19	8	0	0	0	20.76±0.004	20.21	0.55	4.51±0.005	4.60	-0.09
20	12	0	0	0	19.02±0.006	20.21	-1.19	4.61±0.009	4.60	0.01

Analysis of Variance (ANOVA) for L-asparaginase and L-glutaminase activities

Analysis of Variance (ANOVA) for enzyme activities using Design-Expert7® software was performed to confirm the suitability of the model and the results are shown in table 4.11&4.12. A computed F value of 57.75 and 37.76 for the quadratic regression models by ANOVA analysis advise that the models are significant and a Prob>F value is <0.0001 which is less than 0.05. A poorer value of coefficient of variation proposes greater consistency of the experimentation, and in the present analysis the attained CV value of 6.76% for L-asparaginase and 7.93% for L-glutaminase activities validates a higher consistency of the trials. The CV of response under trial is denoted by R^2 whose values constantly ranges between 0 and 1; nearer is the value to 1, robust is the statistical model and healthier is the prediction of response(Montgomery and Myers, 1995).

Table 4.11 ANOVA analysis for L-asparaginase activity

Source	Sum of Squares	df	Mean Square	F Value	p-value	Prob > F
Model	629.13	9	69.90	57.75	<0.0001	significant
A-Temp	115.94	1	115.94	95.79	<0.0001	
B-RPM	17.98	1	17.98	14.86	0.0032	
C-pH	186.11	1	186.11	153.76	<0.0001	
AB	1.48	1	1.48	1.22	0.2948	
AC	43.62	1	43.62	36.04	0.0001	
BC	21.52	1	21.52	17.78	0.0018	
A^2	10.53	1	10.53	8.70	0.0146	
B^2	3.61	1	3.61	2.99	0.1147	
C^2	62.87	1	62.87	51.94	<0.0001	
Residual	12.10	10	1.21			
Lack of Fit	9.81	5	1.96	4.27	0.0685	not significant
Pure Error	2.30	5	0.46			
Cor Total	641.23	19				

R-Squared - 0.9811; Adj R-Squared - 0.9641; Pred R-Squared - 0.7908; C.V. % - 6.76
Model F-value of 57.75 implies the model is significant.
Significant model terms: A, B, C, AC, BC, A^2 and C^2

The R^2 for response of L-asparaginase activity is 0.9811 (Table 4.11), signifying that the model can elucidate 98.11% of inconsistency in the response and only 1.89% of the variations for enzyme activity are not described by it. On the other side the R^2 for L-glutaminase activity is 0.9714, which signifies that model can elucidate 97.14% of inconsistency in the response. Value adjusted R^2 rectifies the R^2 value for the sample size and for the quantity of terms in the model. Adj R^2 of 0.9611 for L-asparaginase activity and 0.9457 for L-glutaminase activity are also decent, supporting the significance of the model developed. The significance of distinct variables is assessed based on their P values, with the more significant terms having a lower P value (Table 4.11&4.12). The Prob>F values smaller than 0.05 denote the significant model terms and in case of L-asparaginase A, B, C, AC, BC, A^2 and C^2 were found to be significant whereas A, C, AB, AC, A^2 and B^2 were significant for L-glutaminase activity.

Table 4.12 ANOVA analysis for L-glutaminase activity

Source	Sum of Squares	df	Mean Square	F Value	p-value	Prob > F
Model	30.98	9	3.44	37.76	<0.0001	significant
A-Temp	3.43	1	3.43	37.67	0.0001	
B-RPM	0.35	1	0.35	3.80	0.0800	
C-pH	8.01	1	8.01	87.87	<0.0001	
AB	6.16	1	6.16	67.58	<0.0001	
AC	1.81	1	1.81	19.80	0.0012	
BC	0.026	1	0.026	0.29	0.6019	
A^2	4.09	1	4.09	44.83	<0.0001	
B^2	0.95	1	0.95	10.47	0.0089	
C^2	0.15	1	0.15	1.68	0.2242	
Residual	0.91	10	0.091			
Lack of Fit	0.51	5	0.10	1.28	0.3980	not significant
Pure Error	0.40	5	0.080			
Cor Total	31.89	19				

R-Squared - 0.9714; Adj R-Squared - 0.9457; Pred R-Squared - 0.8551; C.V. % - 7.93.
Model F-value of 37.76 implies the model is significant.
Significant model terms: A, C, AB, AC, A^2 and B^2

Mutual effect of process variables on enzyme activities was represented in Fig. 4.45. The highest asparaginase activity was attained at Incubation Temperature between 25°C to 40°C, rpm between 100 to 200 and pH from 6 to 8. The 3D surface plots of shared effects of Temperature, rpm and pH on both enzyme activities are shown in Fig. 4.45 (a to c for L- asparaginase activity and d to f for L- glutaminase activity).

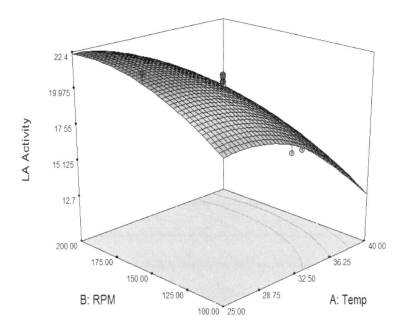

Fig. 4.45a Mutual effect of rpm and temperature on L-asparaginase activity

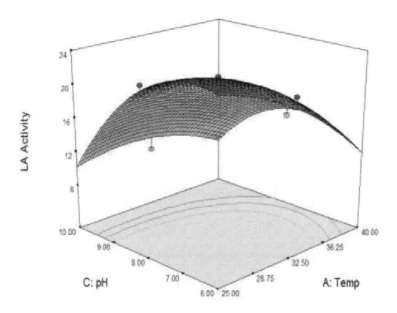

Fig. 4.45b Mutual effect of pH and temperature on L-asparaginase activity

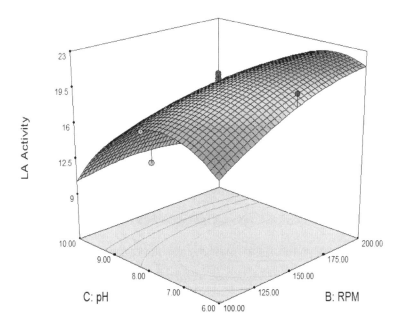

Fig. 4.45c Mutual effect of pH and rpm on L-asparaginase activity

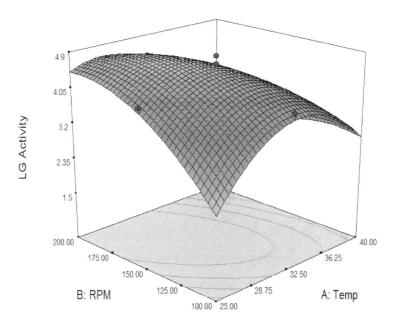

Fig. 4.45d Mutual effect of rpm and temperature on L-glutaminase activity

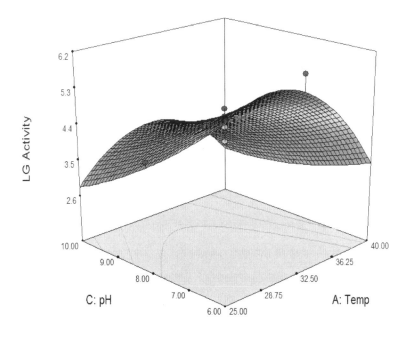

Fig. 4.45e Mutual effect of pH and temperature on L-glutaminase activity

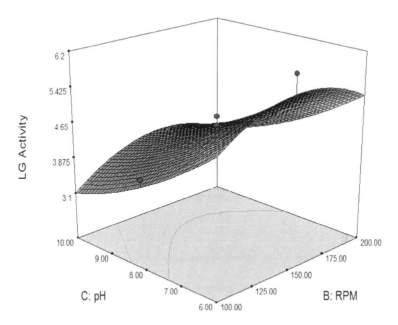

Fig. 4.45f Mutual effect of pH and rpm on L-glutaminase activity

RSM Model validation

RSM proposed ten possible solutions for the L-asparaginase production. In order to validate the RSM model, experiments were conducted at all the ten given solutions and an enzyme activity of 19.02 IUml⁻¹ was achieved which is quite closer to predicted activity of 18.84IUml⁻¹ at the predicted process variable values (Temperature 25°C, pH is 7.96, and rpm 100). Similarly observed L-glutaminase activity of 2.59 IUml⁻¹ is also quite closer to the predicted activity value of 2.34 IUml⁻¹. As the results attained from the experiment were comparable to the model predicted and therefore the RSM model is validated.

Enterobacter aerogenes MTCC 111 resulted in an enzyme activity of 19.02IUml⁻¹after statistical optimization. This is comparatively superior to other organisms producing L-asparaginase enzyme both in submerged as well as solid substrate fermentation. Mukherjee et al., 2000 (Mukherjee, et al., 2000) reported maximum activity of only 0.57 IUml⁻¹ using *Enterobacter aerogenes NCIM 2340* without the addition of L-asparagine. On the other hand *Enterobacter aerogenes* achieved only 18.72 IUml^{-1} of enzyme even after the ANN coupled GA optimization Bhaskar et al., 2011 (Baskar, et al., 2011). They have attained 19.13IUml^{-1}of activity using *Enterobacter aerogenes MTCC 2823* with 1.88% of Tri Sodium Citrate as carbon source. In contrast *Enterobacter aerogenes MTCC 111* is consuming only 1.34% of Tri Sodium Citrate as carbon source (Baskar, et al., 2009). *Pectobacterium carotovorum* gave a maximum L-asparaginase activity of 19.33 IUml^{-1} with 3.5 g/l Lactose as carbon source with a high concentration of L-asparagine 4% in the medium (Singhal and Swaroop, 2013) after a two level statistical optimization (Plackett-Burman Design followed by RSM), where L-asparagine concentration is only 2% in case of *Enterobacter aerogenes MTCC 111*.

Bacillus sp. RKS 20 gave 15.1 IUml^{-1} L-asparaginase activity using with a medium having 1.4% of L-asparagine (Mahajan, et al., 2014). In the present investigation *Enterobacter aerogenes MTCC 111* gave considerably high levels of L-asparaginase activity value using statistical RSM optimization than *E. coli ATCC 11303*with 1.03 IUml^{-1} (Kenari, et al., 2011), *Pseudomonas plecoglossicida RS1* (Shakambari, et al., 2015) with 3.65 IUml^{-1}, *Cladosporium sp.* with 3.74 IUml^{-1} (Kumar, et al., 2013), *Pectobacterium carotovorum MTCC 1428* with 3.74 IUml^{-1} (Sanjeeviroyar, et al., 2010) and *Erwinia aroideae* with1.41 IUml^{-1}activity (Liu and Zajic, 1972). It took even 120 h of fermentation time for some organisms to attain these low levels of enzyme activities. As per our knowledge this is the first report describing the optimum process parameters for L-asparaginase production by *Enterobacter aerogenes MTCC 111* using statistical optimization with less incubation time of 30 h in shake flask fermentation, which is very important in industrial point of view.

c. Development of Neural Network model and result analysis

With three inputs and two outputs using feed forward back propagation network and TRAINLM training function; training, testing and validation of NN were carried out. Results were shown in table 4.13&4.14. The outcomes found from the analysis were awfully pleasing, and an elevated regression value of 0.99476 was attained. Based on data training, testing and its validation, performance curve was gained (Fig. 4.46) using MATLAB 2009a. Regression plot showing the output vs.

118

target was attained with ten hidden nodes and a regression value of 0.99476 was accomplished which indicates model validation. Table 4.13 shows experimental and predicted data from statistical regression and ANN.

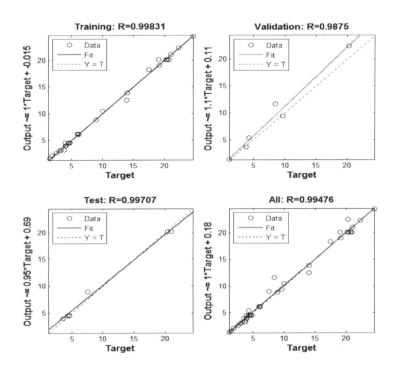

Fig. 4.46 Output vs. target regression plot

Table 4.13 Comparison of values for L-asparaginase activity

S.No.	Temp (°C)	Rpm	pH	Experimental values (IUml⁻¹)	RSM predicted values (IUml⁻¹)	NN predicted values (IUml⁻¹)
1	-1	-1	-1	20.97	18.96	21.07
2	+1	-1	-1	20.45	8.34	20.13
3	-1	+1	-1	22.27	25.79	22.29
4	+1	+1	-1	13.96	13.45	12.51
5	-1	-1	+1	9.61	8.95	9.38
6	+1	-1	+1	20.93	7.67	20.13
7	-1	+1	+1	8.91	9.21	8.84
8	+1	+1	+1	20.76	6.21	20.13
9	-1	0	0	20.32	21.65	22.50
10	+1	0	0	10.05	14.84	10.48
11	0	-1	0	13.98	17.72	13.85
12	0	+1	0	19.02	20.40	20.13
13	0	0	-1	20.19	19.74	20.13
14	0	0	+1	17.53	11.11	18.29
15	0	0	0	6.09	20.21	6.16
16	0	0	0	8.48	20.21	11.64
17	0	0	0	19.15	20.21	19.01
18	0	0	0	7.57	20.21	8.98
19	0	0	0	24.61	20.21	24.44
20	0	0	0	20.42	20.21	20.13

d. Genetic Algorithm based Multi Objective Process Optimization

The nonlinear statistical regression equation obtained from RSM was optimized using GA and the plausible results were described in table 4.15. For L-asparaginase activity (Objective 1) utmost response (enzyme activity) of 26.87 IUml⁻¹ was achieved at the following optimal conditions, i.e., Incubation time - 30 h, pH - 6.32, Temperature - 25°C, and rpm - 194. At these conditions the predicted maximum enzyme activity is 26.73 IUml⁻¹. On the other hand the L-glutaminase activity (Objective 2) observed was 4.95 IUml⁻¹which was close to predicted value of 5.72 IUml⁻¹. Pareto front chart is given in figure 4.47.

120

Table 4.14 Comparison of values for L-glutaminase activity

S.No.	Temp (°C)	Rpm	pH	Experimental values (IUml⁻¹)	RSM predicted values (IUml⁻¹)	NN predicted values (IUml⁻¹)
1	-1	-1	-1	6.15	6.15	3.97
2	+1	-1	-1	4.81	4.81	3.61
3	-1	+1	-1	4.09	4.09	5.99
4	+1	+1	-1	3.01	3.01	2.11
5	-1	-1	+1	3.86	3.86	1.12
6	+1	-1	+1	4.35	4.35	2.65
7	-1	+1	+1	2.57	2.57	3.36
8	+1	+1	+1	4.51	4.51	1.38
9	-1	0	0	4.39	4.39	3.96
10	+1	0	0	3.29	3.29	2.79
11	0	-1	0	1.89	1.89	3.82
12	0	+1	0	4.61	4.61	4.19
13	0	0	-1	3.99	3.99	5.73
14	0	0	+1	3.97	3.97	3.94
15	0	0	0	1.48	1.48	4.60
16	0	0	0	1.25	1.25	4.60
17	0	0	0	3.79	3.79	4.60
18	0	0	0	3.59	3.59	4.60
19	0	0	0	5.98	5.98	4.60
20	0	0	0	4.61	4.61	4.60

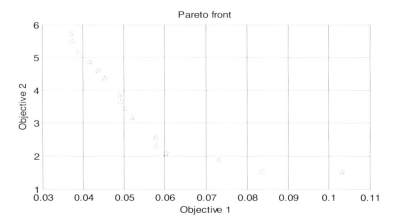

Fig. 4.47 Pareto front for multi objective optimization

Validation of GA outcomes

The results specify that highest L-asparaginase activity was obtained when Temperature, rpm and pH were 25°C, 100, 7.96 and 25°C, 194, 6.32 respectively for RSM and GA. These optimized process parameters were validated by conducting an experiment, and the resulted L-asparaginase enzyme activity of 26.86 IUml⁻¹is quite closer to predicted activity of 26.73 IUml⁻¹in case of GA optimization (Table 4.15). The observed L-glutaminase activity values are also quite closer to the predicted values; however they were a bit high in GA optimization.

The predicted and experimentally determined L-asparaginase activity by *Enterobacter aerogenes MTCC 111* are higher than the activity attained by *Aspergillus terreus, Escherichia coli* and *Pectobacterium carotovorum* (Baskar and Renganathan, 2009; Ghoshoon, et al., 2015; Sanjeeviroyar, et al., 2010). This novel bacterium attained the maximum activity at 25°C unlike the other sources of L-asparaginase resulting greatest activity at higher temperatures (Amena, et al., 2010; Badoei-Dalfard, 2015; El-Bessoumy, et al., 2004; Kamble, et al., 2006; Maladkar, et al., 1992; Narayana, et al., 2008; Sobiś and Mikucki, 1990). The fermentation time is also very less compared to *Actinomycetales bacterium BkSoiiA* as reported by Chitrangada Dash et al.,(Dash, et al., 2016).

The results specify that highest L-asparaginase activity was obtained when temperature was maintained at 25°C, pH at 6.32 and rpm at 194. These optimized process parameters were validated by conducting an experiment, and the resulted enzyme activity of 26.87 IUml^{-1} is quite closer to predicted activity of 26.73 IUml^{-1}. Experimental values were compared with predicted responses by RSM and Neural Network (Table 4.13, Table 4.15 and Fig. 4.48, Fig. 4.49). In the present study, Genetic Algorithm optimization gave improved levels of L-asparaginase and more accurate predicted values were given by Neural Network model compared to RSM prediction.

Table 4.15 Optimized process variables by RSM and GA

Optimization Method	Temp (°C)	Rpm	pH	Observed		Predicted	
				LA activity (IUml⁻¹)	LG activity (IUml⁻¹)	LA activity (IUml⁻¹)	LG activity (IUml⁻¹)
RSM	25	100	7.96	19.02±0.002	2.59±0.001	18.84	2.34
GA	25	194	6.32	26.87±0.015	4.95±0.021	26.73	5.72

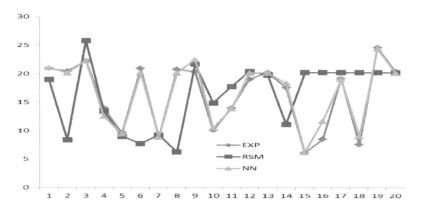

Fig. 4.48 Comparison of values experimental (Exp), RSM predicted (RSM) and
NN predicted (NN) for L-asparaginase activity

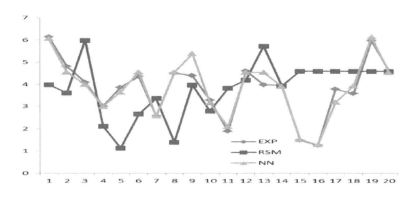

Fig. 4.49 Comparison of values experimental (EXP), RSM predicted (RSM)
and NN predicted (NN) for L-glutaminase activity

4.2.4 Purification of L-asparaginase enzyme

The partial purification of the L-asparaginase crude extract that was affected by the ammonium sulfate (70%) precipitation showed that most of the enzyme activity was preserved in the precipitate. The total protein decreased from 1531.13 to 13.91 mg in the ammonium sulfate precipitation step. The specific activity increased to 440.23 and 985.88 IU/mg after the DEAE Cellulose and Sephadex G75 steps respectively (Table 4.16). Fig. 4.50 shows the profile of the Ion exchange fraction purification on Sephadex G75 gel filtration column chromatography. A sharp distinctive peak of L-asparaginase activity, which fits with only one protein peak, was obtained. Finally 121.49 fold purification of L-asparaginase enzyme purification with 7.38% of yield was achieved. Molecular weight of L-asparaginase SDS-PAGE showed that the enzyme is purified properly and is present only as one band in gel in which lane1-marker (Bangalore Genei), lane2-crude enzyme sample, lane3-Ammonium Sulphate precipitate, lane 4-DEAE Cellulose purified sample and lane5-Sephadex G-75 purified enzyme sample. By using marker with known molecular weights, it was identified that the apparent molecular weight of *Enterobacter aerogenes MTCC 111* L-asparaginase was 37 kDa (Fig. 4.51).

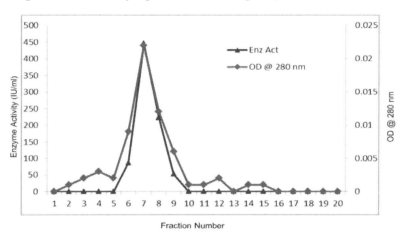

Fig. 4.50 Elution profile of L-asparaginase by sephadex G-75

(All fractions were monitored for total protein at 280 nm and L-asparaginase activity)

Table 4.16 Purification scheme of L-asparaginase from *Enterobacter aerogenes*
MTCC 111

Type of Sample	Volume Collected (ml)	Total Enzyme Activity (IUml⁻¹)	Total Protein (mg/ml)	Specific Activity (IU/mg)	Fold Purification	Yield (%)
Crude	450	12085.34	1531.13	7.89	0.0000	100.00
Ammonium Sulphate Precipitation and Dialysis	15	2404.54	13.91	172.86	21.90	19.90
Ion Exchange Chromatography (DEAE Cellulose)	5	1985.45	4.51	440.23	55.77	16.43
Gel Filtration Chromatography (SephadexG-75)	2	892.22	0.905	985.88	121.48	7.38

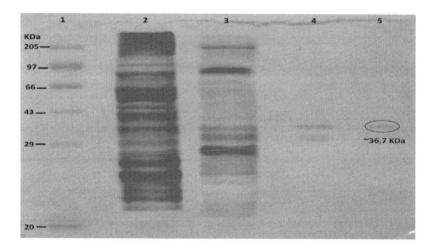

Fig. 4.51 SDS PAGE analysis of L-asparaginase purification

127

4.2.5 Anti-cancer testing of purified L-asparaginase *of Enterobacter aerogenes MTCC 111*

The viability of leukemic cells was tested by standard MTT Assay. Percent of cell viability was decreased with the increased concentration of L-asparaginase from *Enterobacter aerogenes MTCC 111* (Table 4.17, Fig. 4.52, Fig. 4.53). A concentration of 15 IUml^{-1} resulted in 12.13 % of viable cancer cells. Using a micro plate reader, measurement of absorbance was done at 550 nm and the consequences were conveyed as viable cells percentage with respective to control cells.

Table 4.17 Cell viability by MTT assay

Enzyme Concentration	Percentage Cell Viability (%)		
	24 Hours	48 Hours	72 Hours
C1 (2 IUml^{-1})	75.54	74.85	69.85
C2 (5 IUml^{-1})	62.59	58.76	56.82
C3 (10IUml^{-1})	41.29	40.02	34.52
C4 (15 IUml^{-1})	16.14	14.97	12.13

One way ANOVA analysis reveals that the percent cell viability of HL 60 cells is significant (P value = 0.00001) with respective to concentration of enzyme. On the other hand the same cell viability percentage was insignificant (P value = 0.9525) with respective to different time intervals (**ANNEXURE D and E**).

Cell viability (%) was calculated as:

$$\% \text{ Cell Viability } = \frac{\text{Mean absorbance of the sample}}{\text{Mean absorbance of the control}} \text{ X } 100$$

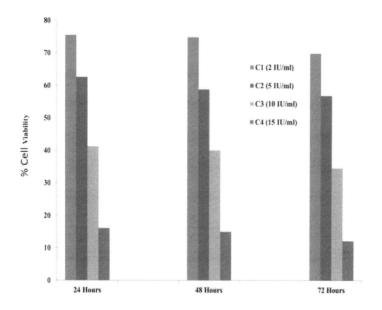

Fig. 4.52 Effect of L-asparaginase on leukemic cells with increasing
concentrations

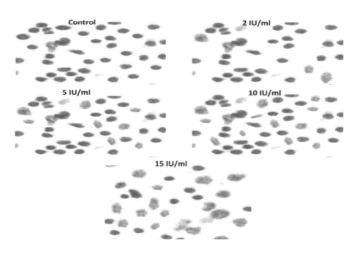

Fig. 4.53 Asparaginase induced morphological changes on leukemic cells

El-Sayed et al. 2011 tested the cytotoxic activity of L-asparaginase from chicken liver on different cancer cell lines like hepatic carcinoma, breast cancer and colon cancer cell lines. The survival rates of cancer lines were decreased with the increased dosage levels of anti-cancer enzyme (El-Sayed, et al., 2011). Efficiency of recombinant enzyme on different cancer cell lines was studied by Guo et al. in 2002. The maximum percentage of cell inhibition was 72 against p815 cell lines (Guo et al. 2002). The antitumor activity of L-asparaginase from *Erwinia carotovora* against different human and animal leukemic and solid tumor cell lines was done by Abakumova et al. in 2012. At 72 hr they obtained 87.5% of cell death. Selvam and Vishnupriya in 2013 achieved only 59.09% of cell death using *Streptomyces acrimycini NGP* L-asparaginase at 70 µg/ml. As, the percent of cell viability of L-asparaginase from *Enterobacter aerogenes KCTC 2190/MTCC 111* at 72 hr is only 12.13, the results are well comparable with the other sources of enzyme drug.

4.2.6 Acrylamide degradation studies

a. Effect of L-asparaginase on materialization of polyacrylamide

Instantaneous acrylamide solidification happened with no L-asparaginase addition. However solidification was hindered by L-asparaginase supplementation in increased concentrations (volume) (Table 4.18).

Table 4.18 Time required for polymerization of acrylamide in the presence of various concentrations of L-asparaginase

S.No	10% Acrylamide solution (ml)	Amount of enzyme (cell free extract) (ml)	Time of solidification of acrylamide (min)
1	5	0.0	Immediate
2	5	0.5	2
3	5	1	5
4	5	3	20
5	5	5	35

Acrylamide is a potential cause of a wide spectrum of toxic effects and is classified as probably "carcinogenic to humans". Asparaginase also find application in food manufacturing companies as it is used to reduce asparagine present in food and thereby, reduce the risk of formation of acrylamide. Acrylamide is formed as a reaction product between asparagine and reducing sugars when certain foods are baked or fried at temperatures exceeding 120°C. Microbial degradation of acrylamide has been explored extensively with a diversity of microbes mainly *Pseudomonas, Bacillus, Arthrobacter, Xanthomonas* and *Rhodococcus* (Cha and Chambliss, 2011). Several acrylamide degraders use a coupling reaction of nitrile hydratase and amidase for biotransformation of acrylonitrile to acrylic acid via acrylamide as an intermediate. Many aerobic microorganisms utilize acrylamide as their sole source of carbon and energy (Wang and Lee, 2007).

L-asparaginase from *Bacillus subtilis KDPS1* resulted in 90-95% degradation of acrylamide in fried potatoes (Sanghvi et al. 2016). *Bacillus licheniformis (RAM-8)* L-asparaginase inhibited poly-acrylamide formation in 10% acrylamide solution and reduced acrylamide formation in fried potatoes by 80% (Mahajan, et al., 2012) where as Pedreschi et al. obtained only 60% of acrylamide degradation (Pedreschi et al. 2008). Kukurová et al. 2009 also attained upto 90% of acrylamide degradation in a fried-dough pastry model (Kukurová et al. 2009). The final level of acrylamide in biscuits and bread was decreased by about 81.6% and 94.2%, respectively, upon treatment with 10 U/mg of flour using L-Asparaginase from *Rhizomucor miehei* (Huang et al. 2014). Though many studies on the effect of L-asparaginase in acrylamide degradation are available, delay in the solidification acrylamide solution upon the addition of enzyme is an alternate method to study the same enzyme effect. In a study carried out by Mahajan, et al., 2012, with the addition of 3 ml of crude enzyme to 10% acrylamide solution; the solidification of solution was delayed by 25 min. Where as with the Enterobacter L-asparagonase in the present study, the results are comparable with *B. licheniformis (RAM-8)* (Mahajan, et al., 2012) and the solidification was further delayed by 35 min with an increased levels of L-asparaginase enzyme.

CONCLUSIONS AND SCOPE FOR FUTURE WORK

5.1 Conclusions

An attempt was made towards identification of new and efficient bacterial source for the production of L-asparaginase for the therapy of ALL by a two layer approach which includes the *In silico* and *In vitro* studies.

1. Screening of *Enterobacter aerogenes KCTC 2190/MTCC* 111was done based on the amino acid sequence provided by Shin et al., 2012 using *In silico* approaches. As the crystal structure of the L-asparaginase from *Enterobacter aerogenes KCTC 2190/MTCC* 111 is not determined, a hypothetical structure was modeled by homology modeling approach using MODELLER 9.1. Comparative studies were done against other type I and type II L-asparaginases from diverse sources by molecular docking approach using different docking tools.

2. As the enzyme has dual activity namely L-asparaginase and L-glutaminase activity, a maximum L-asparaginase and minimal L-glutaminase activity is desired. This was evaluated by testing enzyme's binding affinity against both L-asparaginase and L-glutamine substrates. All the molecular docking results clearly and strongly supported the novel *Enterobacter aerogenes KCTC2 190/MTCC 111* L-Asparaginase as a new therapeutic drug for ALL therapy over the existing ones.

3. MD simulations were performed for the apo enzymes and the docked complexes gained from molecular docking to ratify the stability in dynamic system. Overall result of this computational study confirms the stability of desired *Enterobacter aerogenes KCTC 2190/MTCC 111* L-Asparaginase+L-Asn complex over the L-Asparaginase+L-Gln complex.

4. In MD simulation studies RMSD, RMSF and H-bond analysis strongly supports the novel *Enterobacter aerogenes KCTC 2190/MTCC 111* L-Asparaginase as a new potent therapeutic molecule with similar therapeutic effects for better curing of ALL compared to the existing drugs i.e., Elspar[®] and Erwinaze[®].

5. *In vitro* studies were carried out for the L-asparaginase from *Enterobacter aerogenes KCTC 2190/MTCC 111* which includes the production, optimization, purification, anti-cancer activity testing and acrylamide degradation efficiency of L-asparaginase from novel enzyme. Enzyme production and optimization was done initially by traditional OFAT method in which L-asparaginase activity was enhanced from 2.8 IUml^{-1} to 10.6 IUml^{-1} where a 3.8 fold increase in enzyme activity was attained.

6. Statistical optimization techniques were also employed for further optimization of L-asparaginase activity of *Enterobacter aerogenes KCTC 2190/MTCC 111*. Both physical and chemical variables of bioprocess were screened initially by Plackett-Burman Design. Further RSM was employed to optimize the significant physical variables from Plackett-Burman results by which an activity of 19.02 IUml^{-1} was achieved. This is a 1.9 fold increase in activity compared to OFAT based optimization.

7. Genetic Algorithm based multi objective optimization by was done to increase L-asparaginase and decrease L-glutaminase activity of the novel enzyme. This resulted in an enhancement of L-asparaginase activity to 26.86 IUml^{-1} which is a 2.5 fold improvement comparatively with traditional OFAT optimization.

8. After optimization of culture conditions, enzyme purification was performed by different steps namely, Ammonium Sulphate precipitation & dialysis, ion exchange chromatography (DEAE Cellulose) and gel filtration chromatography (Sephadex G-75). This resulted in an enhanced specific activity of the enzyme from 7.89 IU/mg to 985.88 IU/mg, fold purification of 0 to 121.48 with a final yield of 7.38 %.

133

9. Anti-cancer activity of the purified form of enzyme was tested on Leukemic Cells (HL-60) by standard MTT assay method. Percentage of cell viability was decreased with the increased concentration of L-asparaginase. A concentration of 15 IUml^{-1}*Enterobacter aerogenes KCTC 2190/MTCC 111 L-asparaginase* resulted in 12.13 % of viable cancer cells after 72 h.

10. Acrylamide Degradation by L-asparaginase from *Enterobacter aerogenes MTCC111*was also performed using crude enzyme preparation. Solidification of acrylamide was immediate in absence of enzyme and was delayed by 35 min with the addition of 5ml crude L-asparaginase to acrylamide solution.

5.2 Scope for the future work

Amino acids in the active site of L-asparaginase can be replaced in order to improve the binding efficiency of the enzyme drug with L-asparagine substrate over the L-glutamine substrate. Recombinant/mutant strain of *Enterobacter aerogenes KCTC 2190/MTCC 111* can be tried for the improved production of L-asparaginase. The anti-leukemic drug can also be improved in such a way that it work with higher therapeutic efficacy against different types of other cancers.

BIBLIOGRAPHY

Cancer - Signs and symptoms. NHS Choices. Retrieved 10 June 2014.

Defining Cancer". National Cancer Institute. Retrieved 10June 2014. p.2001-09. In.

Abakumova, O.Y., *et al.* Antitumor activity of L-asparaginase from erwinia carotovora against different human and animal leukemic and solid tumor cell lines, *Biochemistry (Moscow) Supplemental Series B: Biomedical Chemistry* 2012; **6**, 307-316.

Abdel-All, S., *et al.* Studies on the asparaginolytic activity of the brown-pigmented Streptomycetes. *JOURNAL OF DRUG RESEARCH-CAIRO-* 1998;22:171-194.

Abdel-Fattah, Y.R. and Olama, Z.A. L-asparaginase production by Pseudomonas aeruginosa in solid-state culture: evaluation and optimization of culture conditions using factorial designs. *Process Biochemistry* 2002;38(1):115-122.

Abdel, F., Yasser, R. and Olama, Z.A. Studies on the asparaginolytic enzymes of Streptomyces. *Egyptian J Microbiol* 1998;30:155-159.

Aghaiypour, K., Wlodawer, A. and Lubkowski, J. Structural basis for the activity and substrate specificity of Erwinia chrysanthemi L-asparaginase. *Biochemistry* 2001;40(19):5655-5664.

Alder, B. and Wainwright, T. Phase transition for a hard sphere system. *The Journal of chemical physics* 1957;27(5):1208.

Alder, B.J. and Wainwright, T. Studies in molecular dynamics. I. General method. *The Journal of Chemical Physics* 1959;31(2):459-466.

Alexanian, R., *et al.* Treatment for multiple myeloma: combination chemotherapy with different melphalan dose regimens. *Jama* 1969;208(9):1680-1685.

Almeida, J.S. Predictive non-linear modeling of complex data by artificial neural networks. *Current opinion in biotechnology* 2002;13(1):72-76.

Amena, S., *et al.* Production, purification and characterization of L-asparaginase from Streptomyces gulbargensis. *Brazilian journal of Microbiology* 2010;41(1):173-178.

Anand, P., *et al.* Cancer is a preventable disease that requires major lifestyle changes. *Pharmaceutical research* 2008;25(9):2097-2116.

Andrew, M., Brooker, L. and Mitchell, L. Acquired antithrombin III deficiency secondary to asparaginase therapy in childhood acute lymphoblastic leukaemia. *Blood Coagulation & Fibrinolysis* 1994;5(1):S24-36.

Andrusier, N., Nussinov, R. and Wolfson, H.J. FireDock: fast interaction refinement in molecular docking. *Proteins: Structure, Function, and Bioinformatics* 2007;69(1):139-159.

Arcaklıoğlu, E., Çavuşoğlu, A. and Erişen, A. Thermodynamic analyses of refrigerant mixtures using artificial neural networks. *Applied Energy* 2004;78(2):219-230.

Aricò, M., *et al.* Familial clustering of Langerhans cell histiocytosis. *British journal of haematology* 1999;107(4):883-888.

Arrivukkarasan, S., *et al.* Effect of medium composition and kinetic studies on extracellular and intracellular production of L-asparaginase from Pectobacterium carotovorum. *Food Science and Technology International* 2010;16(2):115-125.

Arya, L. Childhood cancer-challenges and opportunities. *Indian journal of pediatrics* 2003;70(2):159-162.

Aszalos, A. Antitumor Compounds of Natural Origin. In.: Boca Raton, CRC Press; 1982.

Aszalos, A. and Berdy, J. Cytotoxic and antitumor compounds from fermentations. *A. Rep. ferment. Process* 1978;2:305-333.

Backman, K.C. Regulated protein production using site-specific recombination. In.: Google Patents; 1987.

Badoei-Dalfard, A. Purification and characterization of l-asparaginase from Pseudomonas aeruginosa strain SN004: Production optimization by statistical methods. *Biocatalysis and Agricultural Biotechnology* 2015;4(3):388-397.

Bano, M. and Sivaramakrishnan, V. Preparation and properties of L-asparaginase from green chillies (Capsicum annum L.). *Journal of Biosciences* 1980;2(4):291-297.

Bas, D. and Boyaci, I.H. Modeling and optimization II: Comparison of estimation capabilities of response surface methodology with artificial neural networks in a biochemical reaction. *Journal of Food Engineering* 2007;78(3):846-854.

Basha, N.S., *et al.* Production of Extracellular Anti-leukaemic Enzyme Lasparaginase from Marine Actinomycetes by Solidstate and Submerged Fermentation: Purification and Characterisation. *Tropical Journal of Pharmaceutical Research* 2009;8(4).

Baskar, G., *et al.* Optimization of carbon and nitrogen sources for L-asparaginase production by Enterobacter aerogenes using response surface methodology. *Chemical and Biochemical Engineering Quarterly* 2009;23(3):393-397.

Baskar, G., Rajasekar, V. and Renganathan, S. Modeling and Optimization of L-asparaginase Productionby Enterobacter Aerogenes Using Artificial Neural Network Linked Genetic Algorithm. *International Journal of Chemical Engineering and Applications* 2011;2(2):98.

Baskar, G. and Renganathan, S. Production of L-asparaginase from natural substrates by Aspergillus terreus MTCC 1782: Effect of substrate, supplementary nitrogen source and L-asparagine. *International Journal of Chemical Reactor Engineering* 2009;7(1).

Bell, T.L. and Adams, M.A. Ecophysiology of ectomycorrhizal fungi associated with Pinus spp. in low rainfall areas of Western Australia. *Plant Ecology* 2004;171(1-2):35-52.

Benkert, P., Tosatto, S.C. and Schomburg, D. QMEAN: A comprehensive scoring function for model quality assessment. *Proteins: Structure, Function, and Bioinformatics* 2008;71(1):261-277.

Bennett, J., *et al.* Proposals for the classification of chronic (mature) B and T lymphoid leukaemias. French-American-British (FAB) Cooperative Group. *Journal of Clinical Pathology* 1989;42(6):567-584.

Benny, P. Biological properties of l-sparaginase. 2010.

Biesalski, H.K., *et al.* European consensus statement on lung cancer: risk factors and prevention. lung cancer panel. *CA: a cancer journal for clinicians* 1998;48(3):167-176.

Bonneau, R. and Baker, D. Ab initio protein structure prediction: progress and prospects. *Annual review of biophysics and biomolecular structure* 2001;30(1):173-189.

Borek, D., *et al.* Expression, purification and catalytic activity of Lupinus luteus asparagine β-amidohydrolase and its Escherichia coli homolog. *European journal of biochemistry* 2004;271(15):3215-3226.

Borek, D., *et al.* Isolation and characterization of cDNA encoding L-asparaginase from Lupinus luteus. *Plant Physiol* 1999;119:1568-1570.

Borkotaky, B. and Bezbaruah, R. Production and properties of asparaginase from a newErwinia sp. *Folia microbiologica* 2002;47(5):473-476.

Boyd, J.W. and Phillips, A.W. Purification and properties of L-asparaginase from Serratia marcescens. *Journal of bacteriology* 1971;106(2):578-587.

Bruneau, L., Chapman, R. and Marsolais, F. Co-occurrence of both L-asparaginase subtypes in Arabidopsis: At3g16150 encodes a K+-dependent L-asparaginase. *Planta* 2006;224(3):668-679.

Bu" Lock, J., *et al.* Metabolic development and secondary biosynthesis in Penicillium urticae. *Canadian journal of microbiology* 1965;11(5):765-778.

Bussolati, O., *et al.* Characterization of apoptotic phenomena induced by treatment with L-asparaginase in NIH3T3 cells. *Experimental cell research* 1995;220(2):283-291.

Campbell, H., *et al.* Two L-asparaginases from Escherichia coli B. Their Separation, Purification, and Antitumor Activity*. *Biochemistry* 1967;6(3):721-730.

Campbell, H.A. and Mashburn, L.T. L-Asparaginase EC-2 from Escherichia coli. Some substrate specificity characteristics. *Biochemistry* 1969;8(9):3768-3775.

Cannon, M. Encyclopaedia of antibiotics: by John S. Glasby John Wiley and Sons; London, New York, Sydney, Toronto, 1976 372 pages.£ 17.50, $39.00. In.: No longer published by Elsevier; 1978.

Capizzi, R. and Cheng, Y. Therapy of neoplasia with asparaginase. *Enzymes as Drugs* 1981:2-24.

Case, D.A., *et al.* The Amber biomolecular simulation programs. *Journal of computational chemistry* 2005;26(16):1668-1688.

Case, D.A., *et al.* AMBER 9. *University of California, San Francisco* 2006;45.

Cassady, J.M. and Douros, J.D. Anticancer agents based on natural product models. Academic Press; 1980.

Cedar, H. and Schwartz, J.H. Production of L-asparaginase II by Escherichia coli. *Journal of bacteriology* 1968;96(6):2043-2048.

Cha, M. and Chambliss, G.H. Characterization of acrylamidase isolated from a newly isolated acrylamide-utilizing bacterium, Ralstonia eutropha AUM-01, *Current microbiology* 2011, **62**, 671-678.

Chambers, A.F., Groom, A.C. and MacDonald, I.C. Metastasis: dissemination and growth of cancer cells in metastatic sites. *Nature Reviews Cancer* 2002;2(8):563-572.

Chang, C.-Y., Lee, C.-L. and Pan, T.-M. Statistical optimization of medium components for the production of Antrodia cinnamomea AC0623 in submerged cultures. *Applied microbiology and biotechnology* 2006;72(4):654-661.

Chatterjea, M. and Shinde, R. Textbook of Medical Biochemistry Jaypee Brothers Medical Publishers. In.: India; 2002.

Clavell, L.A., *et al.* Four-agent induction and intensive asparaginase therapy for treatment of childhood acute lymphoblastic leukemia. *New England journal of medicine* 1986;315(11):657-663.

Colovos, C. and Yeates, T.O. Verification of protein structures: patterns of nonbonded atomic interactions. *Protein science* 1993;2(9):1511-1519.

Creasey, W.A. Cancer: An introduction. . *1981*. 1981.

Curran, M.P., *et al*. A specific L-asparaginase from Thermus aquaticus. *Archives of biochemistry and biophysics* 1985;241(2):571-576.

Dash, C., Mohapatra, S.B. and Maiti, P.K. Optimization, purification, and characterization of L-asparaginase from Actinomycetales bacterium BkSoiiA. *Preparative Biochemistry and Biotechnology* 2016;46(1):1-7.

Dasu, V.V. and Panda, T. Optimization of microbiological parameters for enhanced griseofulvin production using response surface methodology. *Bioprocess Engineering* 2000;22(1):45-49.

Davidson, L., *et al*. L-Asparaginases from Citrobacter freundii. *Biochimica et Biophysica Acta (BBA)-Enzymology* 1977;480(1):282-294.

Dejong, P.J. L-Asparaginase production by Streptomyces griseus. *Applied microbiology* 1972;23(6):1163-1164.

Demain, A.L. Achievements in microbial technology. *Biotechnology advances* 1990;8(1):291-301.

Derst, C., Henseling, J. and Röhm, K.H. Engineering the substrate specificity of Escherichia coli asparaginase II. Selective reduction of glutaminase activity by amino acid replacements at position 248. *Protein Science* 2000;9(10):2009-2017.

Devlin, T.M. Textbook of biochemistry: with clinical correlations. 2006.

Dhevendaran, K. and Anithakumari, Y. L-asparaginase activity in growing conditions of Streptomyces spp. associated with Therapon jarbua and Villorita cyprinoids of Veli Lake, South India. *Fishery Technology* 2002;39(2).

Doll, R. and Peto, R. The causes of cancer: quantitative estimates of avoidable risks of cancer in the United States today. *Journal of the National Cancer Institute* 1981;66(6):1192-1308.

Douros, J. and Suffness, M. New antitumor substances of natural origin. *Cancer treatment reviews* 1981;8(1):63-87.

DRAINAS, D. and DRAINAS, C. A conductimetric method for assaying asparaginase activity in Aspergillus nidulans. *European journal of biochemistry* 1985;151(3):591-593.

Dubinsky, M.C., *et al.* Pharmacogenomics and metabolite measurement for 6-mercaptopurine therapy in inflammatory bowel disease. *Gastroenterology* 2000;118(4):705-713.

Durrant, J.D. and McCammon, J.A. Molecular dynamics simulations and drug discovery. *BMC biology* 2011;9(1):1.

Duval, M., *et al.* Comparison of Escherichia coli–asparaginase withErwinia-asparaginase in the treatment of childhood lymphoid malignancies: results of a randomized European Organisation for Research and Treatment of Cancer—Children's Leukemia Group phase 3 trial. *Blood* 2002;99(8):2734-2739.

El-Bessoumy, A.A., Sarhan, M. and Mansour, J. Production, isolation, and purification of L-asparaginase from Pseudomonas aeruginosa 50071 using solid-state fermentation. *BMB Reports* 2004;37(4):387-393.

El-Naggar, N., Abdelwahed, N.A. and Darwesh, O.M. Fabrication of biogenic antimicrobial silver nanoparticles by Streptomyces aegyptia NEAE 102 as eco-friendly nanofactory. *Journal of microbiology and biotechnology* 2014;24(4):453-464.

142

El-Naggar, N.E.-A. and Abdelwahed, N.A. Application of statistical experimental design for optimization of silver nanoparticles biosynthesis by a nanofactory Streptomyces viridochromogenes. *Journal of Microbiology* 2014;52(1):53-63.

El-Naggar, N.E.-A., El-Bindary, A.A. and Nour, N.S. Production of Antimicrobial Agent Inhibitory to some Human Pathogenic Multidrug-Resistant Bacteria and Candida albicans by Streptomyces sp NEAE-1. *International Journal of Pharmacology* 2013;9(6):335-347.

El-Naggar, N.E.-A., El-Bindary, A.A. and Nour, N.S. Statistical optimization of process variables for antimicrobial metabolites production by Streptomyces anulatus NEAE-94 against some multidrug-resistant strains. *International Journal of Pharmacology* 2013;9(6):322-334.

El-Naggar, N.E.-A., El-Ewasy, S.M. and El-Shweihy, N.M. of Acute Lymphoblastic Leukemia: The Pros and Cons. *International Journal of Pharmacology* 2014;10(4):182-199.

El-Sayed, M., *et al.* Purification, Characterization and Antitumor Activity of L-asparaginase from Chicken liver. *Journal of American Science* 2011;7(1).

Erva, R.R., *et al.* Molecular dynamic simulations of Escherichia coli L-asparaginase to illuminate its role in deamination of asparagine and glutamine residues. *3 Biotech* 2016;6(1):1-7.

Ferrara, M.A., *et al.* Asparaginase production by a recombinant Pichia pastoris strain harbouring Saccharomyces cerevisiae ASP3 gene. *Enzyme and microbial technology* 2006;39(7):1457-1463.

Foda, M., Zedan, H. and Hashem, S. Characterization of a novel L-asparaginase produced by Rhodotorula rubra. *Revista latinoamericana de microbiologia* 1979;22(2):87-95.

Futreal, P.A., *et al.* A census of human cancer genes. *Nature Reviews Cancer* 2004;4(3):177-183.

Gajic, O., *et al.* Early identification of patients at risk of acute lung injury: evaluation of lung injury prediction score in a multicenter cohort study. *American journal of respiratory and critical care medicine* 2011;183(4):462-470.

Galetzka, D., *et al.* Monozygotic twins discordant for constitutive BRCA1 promoter methylation, childhood cancer and secondary cancer. *Epigenetics* 2012;7(1):47-54.

Gambardella, A. Science and innovation: The US pharmaceutical industry during the 1980s. Cambridge University Press; 1995.

Geckil, H., *et al.* Enhanced production of acetoin and butanediol in recombinant Enterobacter aerogenes carrying Vitreoscilla hemoglobin gene. *Bioprocess and biosystems engineering* 2004;26(5):325-330.

Geckil, H. and Gencer, S. Production of L-asparaginase in Enterobacter aerogenes expressing Vitreoscilla hemoglobin for efficient oxygen uptake. *Applied microbiology and biotechnology* 2004;63(6):691-697.

Geckil, H., Gencer, S. and Uckun, M. Vitreoscilla hemoglobin expressing Enterobacter aerogenes and Pseudomonas aeruginosa respond differently to carbon catabolite and oxygen repression for production of L-asparaginase, an enzyme used in cancer therapy. *Enzyme and Microbial Technology* 2004;35(2):182-189.

Ghoshoon, M.B., *et al.* Extracellular Production of Recombinant l-Asparaginase II in Escherichia coli: Medium Optimization Using Response Surface Methodology. *International Journal of Peptide Research and Therapeutics* 2015;21(4):487-495.

Godfrin, Y. and Bertrand, Y. L-asparaginase Introduced into Erythrocytes for the Treatment of Leukaemia (ALL). *BioMedES* 2006;1(1):10-13.

Golden, K.J. and Bernlohr, R.W. Nitrogen catabolite repression of the L-asparaginase of Bacillus licheniformis. *Journal of bacteriology* 1985;164(2):938-940.

Graham, M.L. Pegaspargase: a review of clinical studies. *Advanced drug delivery reviews* 2003;55(10):1293-1302.

Granner, D., Mayes, P. and Rodwell, V. Harper's Biochemistry. *Standford2000* 2000.

Gulati, R., Saxena, R. and Gupta, R. A rapid plate assay for screening l-asparaginase producing micro-organisms. *Letters in applied microbiology* 1997;24(1):23-26.

Gunasekaran, S., *et al.* Effect of culture media on growth and L-asparaginase production in Nocardia asteroides. *Biomedical letters* 1995;52(207):197-201.

Guo, Q.-L., Wu, M.-S. and Chen, Z. Comparison of antitumor effect of recombinant L-asparaginase with wild type one in vitro and in vivo, *Acta Pharmacologica Sinica* 2002; 23, 946-951.

Gupta, N., Dash, S.J. and Basak, U.C. L-asparaginases from fungi of Bhitarkanika mangrove ecosystem. *AsPac J. Mol. Biol. Biotech* 2009;17(1):27-30.

Gurunathan, B. and Sahadevan, R. Design of experiments and artificial neural network linked genetic algorithm for modeling and optimization of L-asparaginase production by Aspergillus terreus MTCC 1782. *Biotechnology and Bioprocess Engineering* 2011;16(1):50-58.

Gurunathan, B. and Sahadevan, R. Production of L-asparaginase from natural substrates by Aspergillus terreus MTCC 1782: Optimization of carbon source and operating conditions. *Int. J. of Chem. Reactor Engineering* 2011;9:1-15.

Halgren, T.A. Merck molecular force field. I. Basis, form, scope, parameterization, and performance of MMFF94. *Journal of computational chemistry* 1996;17(5-6):490-519.

Halgren, T.A. Merck molecular force field. II. MMFF94 van der Waals and electrostatic parameters for intermolecular interactions. *Journal of Computational Chemistry* 1996;17(5-6):520-552.

Halgren, T.A. Merck molecular force field. V. Extension of MMFF94 using experimental data, additional computational data, and empirical rules. *Journal of Computational Chemistry* 1996;17(5-6):616-641.

Halgren, T.A. MMFF VI. MMFF94s option for energy minimization studies. *Journal of Computational Chemistry* 1999;20(7):720-729.

Halgren, T.A. MMFF VII. Characterization of MMFF94, MMFF94s, and other widely available force fields for conformational energies and for intermolecular-interaction energies and geometries. *Journal of Computational Chemistry* 1999;20(7):730-748.

Halgren, T.A. and Nachbar, R.B. Merck molecular force field. IV. Conformational energies and geometries for MMFF94. *Journal of Computational Chemistry* 1996;17(5-6):587-615.

Hawkins, D.S., *et al.* Asparaginase pharmacokinetics after intensive polyethylene glycol-conjugated L-asparaginase therapy for children with relapsed acute lymphoblastic leukemia. *Clinical Cancer Research* 2004;10(16):5335-5341.

Heeschen, V., *et al.* Asparagine catabolism in bryophytes: Purification and characterization of two L-asparaginase isoforms from Sphagnum fallax. *Physiologia Plantarum* 1996;97(2):402-410.

Haskell , C.M., *et al.* L-Asparaginase, *New England Journal of Medicine* 1969, **281**, 1028-1034.

Heinemann, B. and Howard, A.J. Production of tumor-inhibitory L-asparaginase by submerged growth of Serratia marcescens. *Applied microbiology* 1969;18(4):550-554.

Hendriksen, H.V., *et al.* Evaluating the potential for enzymatic acrylamide mitigation in a range of food products using an asparaginase from Aspergillus oryzae. *Journal of Agricultural and Food Chemistry* 2009;57(10):4168-4176.

Hess, B., *et al.* LINCS: a linear constraint solver for molecular simulations. *Journal of computational chemistry* 1997;18(12):1463-1472.

Himabindu, M., *et al.* Optimization of critical medium components for the maximal production of gentamicin by Micromonospora echinospora ATCC 15838 using response surface methodology. *Applied Biochemistry and Biotechnology* 2006;134(2):143-154.

Hosamani, R. and Kaliwal, B. Isolation, molecular identification and optimization of fermentation parameters for the production of L-asparaginase, an anticancer agent by Fusarium equiseti. *International Journal of Microbiology Research* 2011;3(2):108.

Hosamani, R. and Kaliwal, B. L-asparaginase an anti-tumor agent production by Fusarium equiseti using solid state fermentation. *International Journal of Drug Discovery* 2011;3(2):88-99.

Howard, J.B. and Carpenter, F.H. L-asparaginase from Erwinia carotovora substrate specificity and enzymatic properties. *Journal of Biological Chemistry* 1972;247(4):1020-1030.

147

Hsu, K.-C., *et al.* iGEMDOCK: a graphical environment of enhancing GEMDOCK using pharmacological interactions and post-screening analysis. *BMC bioinformatics* 2011;12(1):1.

Huang, L., *et al.* Biochemical characterization of a novel L-Asparaginase with low glutaminase activity from Rhizomucor miehei and its application in food safety and leukemia treatment, *Applied and environmental microbiology* 2014, **80**, 1561-1569.

Hymavathi, M., *et al.* Enhancement of L-asparaginase production by isolated Bacillus circulans (MTCC 8574) using response surface methodology. *Applied biochemistry and biotechnology* 2009;159(1):191-198.

Imada, A., *et al.* Asparaginase and glutaminase activities of micro-organisms. *Microbiology* 1973;76(1):85-99.

Jemal, A., *et al.* Global cancer statistics. *CA: a cancer journal for clinicians* 2011;61(2):69-90.

Jones, G.E. Genetic and physiological relationships between L-asparaginase I and asparaginase II in Saccharomyces cerevisiae. *Journal of bacteriology* 1977;130(1):128-130.

Jones, P.A. and Baylin, S.B. The fundamental role of epigenetic events in cancer. *Nature reviews genetics* 2002;3(6):415-428.

Jürgens, H., *et al.* [Clinical experiences with polyethylene glycol-bound E. coli L-asparaginase in patients with multiple recurrences of acute lymphoblastic leukemia]. *Klinische Padiatrie* 1987;200(3):184-189.

Kafkewitz, D. and Goodman, D. L-Asparaginase production by the rumen anaerobe Vibrio succinogenes. *Applied microbiology* 1974;27(1):206-209.

Kaladhar, D. An in vitro callus induction and isolation, identification, virtual screening and docking of drug from convolvulus alsinoides linn against aging diseases. *International Journal of Pharma Medicine and Biological Sciences* 2012;1:92-103.

Kamble, V., *et al.* Purification of L-asparaginase from a bacteria Erwinia carotovora and effect of a dihydropyrimidine derivative on some of its kinetic parameters. *Indian journal of biochemistry and biophysics* 2006;43(6):391.

Kantarjian, H.M., *et al.* Results of treatment with hyper-CVAD, a dose-intensive regimen, in adult acute lymphocytic leukemia. *Journal of Clinical Oncology* 2000;18(3):547-547.

Karin, M. and Greten, F.R. NF-κB: linking inflammation and immunity to cancer development and progression. *Nature Reviews Immunology* 2005;5(10):749-759.

Kasiske, B.L., *et al.* Cancer after kidney transplantation in the United States. *American Journal of Transplantation* 2004;4(6):905-913.

Katchalski-Katzir, E. Immobilized enzymes—learning from past successes and failures. *Trends in biotechnology* 1993;11(11):471-478.

Kattimani, L., *et al.* Immobilization of Streptomyces gulbargensis in Polyurethane Foam: A Promising Technique for L-asparaginase Production on. *Iranian Journal of Biotechnology* 2009;7(4):199-204.

Kellogg, E.W. and Fridovich, I. Superoxide, hydrogen peroxide, and singlet oxygen in lipid peroxidation by a xanthine oxidase system. *Journal of biological chemistry* 1975;250(22):8812-8817.

Kenari, S.L.D., Alemzadeh, I. and Maghsodi, V. Production of L-asparaginase from Escherichia coli ATCC 11303: Optimization by response surface methodology. *Food and Bioproducts Processing* 2011;89(4):315-321.

Khamna, S., Yokota, A. and Lumyong, S. L-asparaginase production by actinomycetes isolated from some Thai medicinal plant rhizosphere soils. *International Journal of Integrative Biology* 2009;6(1):22-26.

Khushoo, A., Pal, Y. and Mukherjee, K. Optimization of extracellular production of recombinant asparaginase in Escherichia coli in shake-flask and bioreactor. *Applied Microbiology and Biotechnology* 2005;68(2):189-197.

King, M.-C., Marks, J.H. and Mandell, J.B. Breast and ovarian cancer risks due to inherited mutations in BRCA1 and BRCA2. *Science* 2003;302(5645):643-646.

Klumper, E., *et al.* In vitro cellular drug resistance in children with relapsed/refractory acute lymphoblastic leukemia. *Blood* 1995;86(10):3861-3868.

Koeller, K.M. and Wong, C.-H. Enzymes for chemical synthesis. *Nature* 2001;409(6817):232-240.

Kotzia, G.A. and Labrou, N.E. Cloning, expression and characterisation of Erwinia carotovoral-asparaginase. *Journal of biotechnology* 2005;119(4):309-323.

Kotzia, G.A. and Labrou, N.E. L-Asparaginase from Erwinia chrysanthemi 3937: cloning, expression and characterization. *Journal of biotechnology* 2007;127(4):657-669.

Kozak, M., *et al.* Crystallization and preliminary crystallographic studies of five crystal forms of Escherichia coli L-asparaginase II (Asp90Glu mutant). *Acta Crystallographica Section D: Biological Crystallography* 2002;58(1):130-132.

Krasotkina, J., *et al.* One-step purification and kinetic properties of the recombinant l-asparaginase from Erwinia carotovora. *Biotechnology and applied biochemistry* 2004;39(2):215-221.

Kukurová, K., *et al.* Effect of L-asparaginase on acrylamide mitigation in a fried-dough pastry model, *Molecular Nutrition & Food Research* 2009, **53**, 1532-1539.

Kumar, K., Punia, S.S. and Verma, N. Media optimization for the production of anti-leukemic enzyme L-asparaginase from E. coli K-12. *Annals of Biological Research* 2012;3(10):4828-4837.

Kumar, N.M., Ramasamy, R. and Manonmani, H. Production and optimization of L-asparaginase from Cladosporium sp. using agricultural residues in solid state fermentation. *Industrial Crops and Products* 2013;43:150-158.

Kumar, S., Dasu, V.V. and Pakshirajan, K. Localization and production of novel L-asparaginase from Pectobacterium carotovorum MTCC 1428. *Process Biochemistry* 2010;45(2):223-229.

Kumar, S., Dasu, V.V. and Pakshirajan, K. Purification and characterization of glutaminase-free L-asparaginase from Pectobacterium carotovorum MTCC 1428. *Bioresource technology* 2011;102(2):2077-2082.

Kumar, S., Pakshirajan, K. and Dasu, V.V. Development of medium for enhanced production of glutaminase-free L-asparaginase from Pectobacterium carotovorum MTCC 1428. *Applied microbiology and biotechnology* 2009;84(3):477-486.

Lacoste, L., Chaudhary, K.D. and Lapointe, J. Derepression of the glutamine synthetase in neuroblastoma cells at low concentrations of glutamine. *Journal of neurochemistry* 1982;39(1):78-85.

Land, V.J., *et al.* Toxicity of L-asparaginase in children with advanced leukemia. *Cancer* 1972;30(2):339-347.

Lawrenz, M., *et al.* Effects of biomolecular flexibility on alchemical calculations of absolute binding free energies. *Journal of chemical theory and computation* 2011;7(7):2224-2232.

Lemieux, R., *et al.* The conformations of oligosaccharides related to the ABH and Lewis human blood group determinants. *Canadian Journal of Chemistry* 1980;58(6):631-653.

Lindahl, E., Hess, B. and Van Der Spoel, D. GROMACS 3.0: a package for molecular simulation and trajectory analysis. *Molecular modeling annual* 2001;7(8):306-317.

Lins, R.D. and Hünenberger, P.H. A new GROMOS force field for hexopyranose-based carbohydrates. *Journal of computational chemistry* 2005;26(13):1400-1412.

Liu, F. and Zajic, J. L-Asparaginase synthesis by Erwinia aroideae. *Applied microbiology* 1972;23(3):667-668.

Liu, F. and Zajic, J. Fermentation kinetics and continuous process of L-asparaginase production. *Applied microbiology* 1973;25(1):92-96.

Long, S., *et al.* Amino acid residues adjacent to the catalytic cavity of tetramer l-asparaginase II contribute significantly to its catalytic efficiency and thermostability. *Enzyme and microbial technology* 2016;82:15-22.

Lough, T.J., *et al.* L-Asparaginase from developing seeds of Lupinus arboreus. *Phytochemistry* 1992;31(5):1519-1527.

Lovell, S.C., *et al.* Structure validation by Cα geometry: ϕ, ψ and Cβ deviation. *Proteins: Structure, Function, and Bioinformatics* 2003;50(3):437-450.

Lowry, O.H., *et al.* Protein measurement with the Folin phenol reagent. *J biol Chem* 1951;193(1):265-275.

Lubkowski, J., *et al.* Crystal Structure and Amino Acid Sequence of Wolinella Succinogenesl-Asparaginase. *European Journal of Biochemistry* 1996;241(1):201-207.

Lubkowski, J., *et al.* Structural characterization of Pseudomonas 7A glutaminase-asparaginase. *Biochemistry* 1994;33(34):10257-10265.

Lubkowski, J., *et al.* Refined crystal structure of Acinetobacter glutaminasificans glutaminase–asparaginase. *Acta Crystallographica Section D: Biological Crystallography* 1994;50(6):826-832.

MacKerell, A.D., *et al.* CHARMM: the energy function and its parameterization. *Encyclopedia of computational chemistry* 1998.

Mahajan, R.V., *et al.* L-Asparaginase from Bacillus Sp. Rks-20: Process Optimization and Application in the Inhibition of Acrylamide Formation in Fried Foods. *Journal of Proteins & Proteomics* 2014;5(2).

Mahajan, R.V., *et al.* Efficient production of L-asparaginase from Bacillus licheniformis with low-glutaminase activity: optimization, scale up and acrylamide degradation studies. *Bioresource Technology* 2012;125:11-16.

Maladkar, N., Singh, V. and Naik, S. Fermentative production and isolation of L-asparaginase from Erwinia carotovora, EC-113. *Hindustan antibiotics bulletin* 1992;35(1-2):77-86.

Manna, S., *et al.* Purification, characterization and antitumor activity of L-asparaginase isolated from Pseudomonas stutzeri MB-405. *Current Microbiology* 1995;30(5):291-298.

Mashburn, L.T. and Wriston, J.C. Tumor inhibitory effect of L-asparaginase from Escherichia coli. *Archives of Biochemistry and Biophysics* 1964;105(2):450-453.

153

Mashiach, E., et al. FireDock: a web server for fast interaction refinement in molecular docking. *Nucleic acids research* 2008;36(suppl 2):W229-W232.

Matias da Rocha Neto, A. Investigação do comportamento dinâmico de biorreatores contínuos do tipo tanque perfeitamente agitados através de diagramas de bifurcação. 2006.

McCammon, J.A., Gelin, B.R. and Karplus, M. Dynamics of folded proteins. *Nature* 1977;267(5612):585-590.

Menck, C.F. and Munford, V. DNA repair diseases: What do they tell us about cancer and aging? *Genetics and molecular biology* 2014;37(1):220-233.

Meyer, B., et al. L-Asparaginase-associated hyperlipidemia with hyperviscosity syndrome in a patient with T-cell lymphoblastic lymphoma. *Annals of oncology* 2003;14(4):658-659.

Miller, H.K. and Balis, M.E. Glutaminase activity of L-asparagine amidohydrolase. *Biochemical pharmacology* 1969;18(9):2225-2232.

Mishra, A. Production of L-asparaginase, an anticancer agent, from Aspergillus niger using agricultural waste in solid state fermentation. *Applied biochemistry and biotechnology* 2006;135(1):33-42.

Mitchell, L., et al. Increased endogenous thrombin generation in children with acute lymphoblastic leukemia: risk of thrombotic complications in L'Asparaginase-induced antithrombin III deficiency. *Blood* 1994;83(2):386-391.

Miyamoto, S. and Kollman, P.A. SETTLE: an analytical version of the SHAKE and RATTLE algorithm for rigid water models. *Journal of computational chemistry* 1992;13(8):952-962.

Mohapatra, B., Bapuji, M. and Banerjee, U. Production and properties of L-asparaginase from Mucor species associated with a marine sponge (Spirastrella sp.). *Cytobios* 1996;92(370-371):165-173.

Moharam, M., Gamal-Eldeen, A. and El-Sayed, S. Production, immobilization and anti-tumor activity of L-asparaginase of Bacillus sp R36. *J. Am. Sci* 2010;6(8):131-140.

Montgomery, D.C. and Myers, R.H. Response surface methodology: process and product optimization using designed experiments. *Raymond H. Meyers and Douglas C. Montgomery. A Wiley-Interscience Publications* 1995.

Mostafa, S. and Salama, M. L-Asparaginase-producing Streptomyces from the soil of Kuwait. *Zentralblatt für Bakteriologie, Parasitenkunde, Infektionskrankheiten und Hygiene. Zweite Naturwissenschaftliche Abteilung: Mikrobiologie der Landwirtschaft, der Technologie und des Umweltschutzes* 1979;134(4):325-334.

Mukherjee, J., Majumdar, S. and Scheper, T. Studies on nutritional and oxygen requirements for production of L-asparaginase by Enterobacter aerogenes. *Applied Microbiology and Biotechnology* 2000;53(2):180-184.

Nakahama, K., *et al.* Formation of L-asparaginase by Fusarium species. *Microbiology* 1973;75(2):269-273.

Nandy, P., Periclou, A. and Avramis, V. The synergism of 6-mercaptopurine plus cytosine arabinoside followed by PEG-asparaginase in human leukemia cell lines (CCRF/CEM/0 and (CCRF/CEM/ara-C/7A) is due to increased cellular apoptosis. *Anticancer research* 1997;18(2A):727-737.

Narayana, K., Kumar, K. and Vijayalakshmi, M. L-asparaginase production by Streptomyces albidoflavus. *Indian Journal of Microbiology* 2008;48(3):331-336.

Nawani, N. and Kapadnis, B. Optimization of chitinase production using statistics based experimental designs. *Process biochemistry* 2005;40(2):651-660.

Nester, E.W., *et al.* Crown gall: a molecular and physiological analysis. *Annual Review of Plant Physiology* 1984;35(1):387-413.

Neuman, R.E. and McCoy, T.A. Dual requirement of Walker carcinosarcoma 256 in vitro for asparagine and glutamine. In.: American Association for the Advancement of Science; 1956.

Nilolaev, A., *et al.* [Isolation and properties of a homogeneous L-asparaginase preparation from Pseudomonas fluorescens AG]. *Biokhimiia (Moscow, Russia)* 1974;40(5):984-989.

Norberg, J. and Nilsson, L. Advances in biomolecular simulations: methodology and recent applications. *Quarterly reviews of biophysics* 2003;36(03):257-306.

Ogawa, J. and Shimizu, S. Microbial enzymes: new industrial applications from traditional screening methods. *Trends in Biotechnology* 1999;17(1):13-20.

Ohnuma, T., *et al.* Biochemical and pharmacological studies with asparaginase in man. *Cancer Research* 1970;30(9):2297-2305.

Ollenschläger, G., *et al.* Asparaginase-induced derangements of glutamine metabolism: the pathogenetic basis for some drug-related side-effects. *European journal of clinical investigation* 1988;18(5):512-516.

Organization, W.H. Cancer control: knowledge into action: WHO guide for effective programmes. World Health Organization; 2007.

Organization, W.H. Cancer Fact sheet N 297. February 2014. In.: Retrieved; 2014.

Oza, V.P., *et al.* Anticancer properties of highly purified L-asparaginase from Withania somnifera L. against acute lymphoblastic leukemia. *Applied biochemistry and biotechnology* 2010;160(6):1833-1840.

Park, S., *et al.* Free energy calculation from steered molecular dynamics simulations using Jarzynski's equality. *The Journal of chemical physics* 2003;119(6):3559-3566.

Pastuszak, I. and Szymona, M. Purification and properties of L-asparaginase from Mycobacterium phlei. *Acta Biochimica Polonica* 1975;23(1):37-44.

Patro, K.K.R., Satpathy, S. and Gupta, N. Evaluation of some fungi for L-asparaginase production. *Indian Journal of Fundamental and Applied Life Sciences* 2011;1(4):219-221.

Paul, J. Isolation and characterization of a Chlamydomonas L-asparaginase. *Biochemical Journal* 1982;203(1):109-115.

Paymaster, J. and Gangadharan, P. Mortality from cancer in Greater Bombay. *Journal of the Indian Medical Association* 1971;57(2):63-69.

Pedreschi, F., Kaack, K. and Granby, K. The effect of asparaginase on acrylamide formation in French fries. *Food chemistry* 2008;109(2):386-392.

Peterson, R. and Ciegler, A. L-asparaginase production by Erwinia aroideae. *Applied microbiology* 1969;18(1):64-67.

Phillips, J.C., *et al.* Scalable molecular dynamics with NAMD. *Journal of computational chemistry* 2005;26(16):1781-1802.

Pinheiro, I., *et al.* Production of L-asparaginase by Zymomonas mobilis strain CP4. *Biomaterial and Diagnostic BD* 2001;6:243-244.

Plackett, R.L. and Burman, J.P. The design of optimum multifactorial experiments. *Biometrika* 1946;33(4):305-325.

Potenza, L., *et al.* A t (11; 20)(p15; q11) may identify a subset of nontherapy-related acute myelocytic leukemia. *Cancer genetics and cytogenetics* 2004;149(2):164-168.

Prager, M.D. and Bachynsky, N. Asparagine synthetase in asparaginase resistant and susceptible mouse lymphomas. *Biochemical and biophysical research communications* 1968;31(1):43-47.

Prakasham, R., *et al.* l-asparaginase production by isolated Staphylococcus sp.–6A: design of experiment considering interaction effect for process parameter optimization. *Journal of applied microbiology* 2007;102(5):1382-1391.

Pratt, D., Hudson, B. and Hudson, B. Food antioxidants. *London: Elsevier Applied Science* 1990:171-191.

Pritsa, A., *et al.* Antitumor activity of L-asparaginase from Thermus thermophilus. *Anti-cancer drugs* 2001;12(2):137-142.

Pritsa, A.A. and Kyriakidis, D.A. L-asparaginase of Thermus thermophilus: Purification, properties and identificaation of essential amino acids for its catalytic activity. *Molecular and cellular biochemistry* 2001;216(1-2):93-101.

Pui, C.-H. Childhood leukemias. *New England Journal of Medicine* 1995;332(24):1618-1630.

Pui, C.-H., Robison, L.L. and Look, A.T. Acute lymphoblastic leukaemia. *The Lancet* 2008;371(9617):1030-1043.

Raha, S., *et al.* Purification and properties of an L-asparaginase from Cylindrocarpon obtusisporum MB-10. *Biochemistry international* 1990;21(6):987-1000.

Rajulapati, S.B., Narasu, L. and Vundavilli, P. Optimization of α-amylase production from Aspergillus Niger using spoiled starch rich vegetables by response surface methodology and Genetic Algorithm. In, *2011 Annual IEEE India Conference*. IEEE; 2011. p. 1-9.

Ramaiah, N. and Chandramohan, D. Production of L-asparaginase by the marine luminous bacteria. 1992.

Ramakrishnan, M. and Joseph, R. Characterization of an extracellular asparaginase of Rhodosporidium toruloides CBS14 exhibiting unique physicochemical properties. *Canadian Journal of microbiology* 1996;42(4):316-325.

Ramya, L., *et al*. In silico engineering of L-asparaginase to have reduced glutaminase side activity for effective treatment of acute lymphoblastic leukemia. *Journal of pediatric hematology/oncology* 2011;33(8):617-621.

Ramya, L., *et al*. L-Asparaginase as potent anti-leukemic agent and its significance of having reduced glutaminase side activity for better treatment of acute lymphoblastic leukaemia. *Applied biochemistry and biotechnology* 2012;167(8):2144-2159.

Reddy, E.R., *et al*. Exploration of the binding modes of l-asparaginase complexed with its amino acid substrates by molecular docking, dynamics and simulation. *3 Biotech* 2016;6(1):1-8.

Reddy, L., *et al*. Optimization of alkaline protease production by batch culture of Bacillus sp. RKY3 through Plackett–Burman and response surface methodological approaches. *Bioresource technology* 2008;99(7):2242-2249.

Reinert, R.B., *et al*. Role of glutamine depletion in directing tissue-specific nutrient stress responses to L-asparaginase. *Journal of Biological Chemistry* 2006;281(42):31222-31233.

159

Reya, T., *et al.* Stem cells, cancer, and cancer stem cells. *nature* 2001;414(6859):105-111.

Reynolds, D. and Taylor, J. The fungal holomorph: A consideration of mitotic meiotic and pleomorphic speciation. *CAB International, Wallingford, UK* 1993.

Ritchie, D.W. Evaluation of protein docking predictions using Hex 3.1 in CAPRI rounds 1 and 2. *Proteins: Structure, Function, and Bioinformatics* 2003;52(1):98-106.

Ritchie, D.W. and Kemp, G.J. Protein docking using spherical polar Fourier correlations. *Proteins: Structure, Function, and Bioinformatics* 2000;39(2):178-194.

Roberts, J., Holcenberg, J.S. and Dolowy, W.C. Isolation, Crystallization, and Properties of Achromobacteraceae Glutaminase-Asparaginase with Antitumor Activity. *Journal of Biological Chemistry* 1972;247(1):84-90.

Rossi, F., Incorvaia, C. and Mauro, M. [Hypersensitivity reactions to chemotherapeutic antineoplastic agents]. *Recenti progressi in medicina* 2004;95(10):476-481.

Rudman, D., *et al.* Observations on the plasma amino acids of patients with acute leukemia. *Cancer research* 1971;31(8):1159-1165.

Sabu, A. Sources, properties and applications of microbial therapeutic enzymes. *Indian Journal of Biotechnology* 2003;2(3):334-341.

Sahoo, S. and Hart, J. Histopathological features of L-asparaginase-induced liver disease. In, *Seminars in liver disease*. Copyright© 2003 by Thieme Medical Publishers, Inc., 333 Seventh Avenue, New York, NY 10001, USA. Tel.:+ 1 (212) 584-4662; 2003. p. 295-300.

160

Sahu, M.K., *et al.* Partial purification and anti-leukemic activity of L-asparaginase enzyme of the actinomycete strain LA-29 isolated from the estuarine fish, Mugil cephalus (Linn.). *Journal of Environmental Biology* 2007;28(3):645.

Šali, A. and Blundell, T.L. Comparative Protein Modelling by Satisfaction of Spatial Restraints. *Journal of Molecular Biology* 1993;234(3):779-815.

Sanches, M., Krauchenco, S. and Polikarpov, I. Structure, substrate complexation and reaction mechanism of bacterial asparaginases. *Current Chemical Biology* 2007;1(1):75-86.

Sanghvi, G., *et al.* Mitigation of acrylamide by l-asparaginase from Bacillus subtilis KDPS1 and analysis of degradation products by HPLC and HPTLC, *SpringerPlus* 2016, **5**, 533.

Sanjeeviroyar, A., *et al.* Sequential optimization and kinetic modeling of L-asparaginase production by Pectobacterium carotovorum in submerged fermentation. *Asia-Pacific Journal of Chemical Engineering* 2010;5(5):743-755.

Sasco, A., Secretan, M. and Straif, K. Tobacco smoking and cancer: a brief review of recent epidemiological evidence. *Lung cancer* 2004;45:S3-S9.

Saviola, A., *et al.* Myocardial ischemia in a patient with acute lymphoblastic leukemia during l-asparaginase therapy. *European journal of haematology* 2004;72(1):71-72.

Savitri, A.N. and Azmi, W. Microbial L-asparaginase: A potent antitumour enzyme. *Indian Journal of Biotechnology* 2003;2(2):184-194.

Schneidman-Duhovny, D., *et al.* Nucleic Acids Res 33. *Web Server issue), W363-367* 2005.

Schuèttelkopf, A.W. and Van Aalten, D.M. PRODRG: a tool for high-throughput crystallography of protein–ligand complexes. *Acta Crystallographica Section D: Biological Crystallography* 2004;60(8):1355-1363.

Schwartz, J.H., Reeves, J.Y. and Broome, J.D. Two L-asparaginases from E. coli and their action against tumors. *Proceedings of the National Academy of Sciences* 1966;56(5):1516-1519.

Scott, W.R., *et al.* The GROMOS biomolecular simulation program package. *The Journal of Physical Chemistry A* 1999;103(19):3596-3607.

Selvam, K. and Vishnupriya, B. Partial purification and cytotoxic activity of L-asparaginase from Streptomyces acrimycini NGP. *Int J Res Pharm Biomed Sci* 2013;4:859-869.

Senn, H.M. and Thiel, W. QM/MM methods for biomolecular systems. *Angewandte Chemie International Edition* 2009;48(7):1198-1229.

Shaffer, P.M., *et al.* An asparaginase of Aspergillus nidulans is subject to oxygen repression in addition to nitrogen metabolite repression. *Molecular and General Genetics MGG* 1988;212(2):337-341.

Shakambari, G., *et al.* Industrial effluent as a substrate for glutaminase free l-asparaginase production from Pseudomonas plecoglossicida strain RS1; media optimization, enzyme purification and its characterization. *RSC Advances* 2015;5(60):48729-48738.

Shapiro, R.S., *et al.* Epstein-Barr virus associated B cell lymphoproliferative disorders following bone marrow transplantation. *Blood* 1988;71(5):1234-1243.

Sharma, A. and Husain, I. Optimization of medium components for extracellular glutaminase free asparaginase from Enterobacter cloacae. *Int J Curr Microbiol App Sci* 2015;4(1):296-309.

Sharma, D. and Satyanarayana, T. A marked enhancement in the production of a highly alkaline and thermostable pectinase by Bacillus pumilus dcsr1 in submerged fermentation by using statistical methods. *Bioresource Technology* 2006;97(5):727-733.

Shimizu, S., *et al.* Screening of novel microbial enzymes for the production of biologically and chemically useful compounds. In, *New Enzymes for Organic Synthesis*. Springer; 1997. p. 45-87.

Shin, S.H., *et al.* Complete genome sequence of Enterobacter aerogenes KCTC 2190. *Journal of bacteriology* 2012;194(9):2373-2374.

Shrivastava, A., *et al.* Biotechnological advancement in isolation of anti-neoplastic compounds from natural origin: a novel source of L-asparaginase. *Acta Biomed* 2010;81(2):104-108.

Singhal, B. and Swaroop, K. Optimization of culture variables for the production of L-asparaginase from Pectobacterium carotovorum. *African Journal of Biotechnology* 2013;12(50):6959.

Skibola, C.F., *et al.* Polymorphisms in the methylenetetrahydrofolate reductase gene are associated with susceptibility to acute leukemia in adults. *Proceedings of the National Academy of Sciences* 1999;96(22):12810-12815.

Slater, G.W., *et al.* Modeling the separation of macromolecules: a review of current computer simulation methods. *Electrophoresis* 2009;30(5):792-818.

Smith, A., *et al.* Oxford dictionary of biochemistry and molecular biology. Oxford University Press (OUP); 2000.

Smith, G. and Sansom, M. Dynamic properties of Na+ ions in models of ion channels: a molecular dynamics study. *Biophysical journal* 1998;75(6):2767-2782.

Sobiś, M. and Mikucki, J. Staphylococcal L-asparaginase: enzyme kinetics. *Acta Microbiologica Polonica* 1990;40(3-4):143-152.

Song, W.-J., *et al.* Haploinsufficiency of CBFA2 causes familial thrombocytopenia with propensity to develop acute myelogenous leukaemia. *Nature genetics* 1999;23(2):166-175.

Soniyamby, A., *et al.* Isolation, production and anti-tumor activity of L-asparaginase of Penicillium sp. *International Journal of Microbiological Research* 2011;2(1):38-42.

Soria, M.A., Funes, J.L.G. and Garcia, A.F. A simulation study comparing the impact of experimental error on the performance of experimental designs and artificial neural networks used for process screening. *Journal of Industrial Microbiology and Biotechnology* 2004;31(10):469-474.

Sridhar, G., *et al.* Serum butyrylcholinesterase in type 2 diabetes mellitus: a biochemical and bioinformatics approach. *Lipids in health and disease* 2005;4(1):1.

Stecher, A., *et al.* Stability of L-asparaginase: an enzyme used in leukemia treatment. *Pharmaceutica acta helvetiae* 1999;74(1):1-9.

Stegemann, H. SDS-gel electrophoresis in polyacrylamide, merits and limits. In, *Electrokinetic separation methods*. Elsevier Publishers North Holland; 1979. p. 313-336.

Storti, E. and Quaglino, D. Dysmetabolic and neurological complications in leukemic patients treated with L-asparaginase. In, *Experimental and Clinical Effects of L-Asparaginase*. Springer; 1970. p. 344-349.

Story, M.D., *et al.* L-asparaginase kills lymphoma cells by apoptosis. *Cancer chemotherapy and pharmacology* 1993;32(2):129-133.

STRHOL, W. Industrial antibiotics: today and the future. *Biotechnology of antibiotics, 2nd edn. Marcel Dekker, New York* 1997:1-47.

Sutow, W.W., *et al.* L-asparaginase therapy in children with advanced leukemia The Southwest cancer chemotherapy study group. *Cancer* 1971;28(4):819-824.

Takiar, R., Nadayil, D. and Nandakumar, A. Projections of number of cancer cases in India (2010-2020) by cancer groups. *Asian Pac J Cancer Prev* 2010;11(4):1045-1049.

Tardito, S., *et al.* The inhibition of glutamine synthetase sensitizes human sarcoma cells to L-asparaginase. *Cancer chemotherapy and pharmacology* 2007;60(5):751-758.

Theantana, T., Hyde, K. and Lumyong, S. Asparaginase production by endophytic fungi isolated from some Thai medicinal plants. *Kmitl sci. tech. J* 2007;7(1):13-18.

Thomas, X., *et al.* [Therapeutic alternatives to native L-asparaginase in the treatment of adult acute lymphoblastic leukemia]. *Bulletin du cancer* 2010;97(9):1105-1117.

Tischkowitz, M. and Rosser, E. Inherited cancer in children: practical/ethical problems and challenges. *European Journal of Cancer* 2004;40(16):2459-2470.

Tosa, T., *et al.* L-Asparaginase from Proteus vulgaris. *Applied microbiology* 1971;22(3):387-392.

Tosa, T., *et al.* L-Asparaginase form Proteus vulgaris. Purification, crystallization, and enzymic properties. *Biochemistry* 1972;11(2):217-222.

Tower, D.B., Peters, E.L. and Curtis, W.C. Guinea pig serum L-asparaginase properties, purification, and application to determination of asparagine in biological samples. *Journal of Biological Chemistry* 1963;238(3):983-993.

Triantafillou, D., Georgatsos, J. and Kyriakidis, D. Purification and properties of a membrane-bound L-asparaginase of Tetrahymena pyriformis. *Molecular and cellular Biochemistry* 1988;81(1):43-51.

Trivedi, C.D. and Pitchumoni, C. Drug-induced pancreatitis: an update. *Journal of clinical gastroenterology* 2005;39(8):709-716.

Uggeri, J., *et al.* Suppression of anionic amino acid transport impairs the maintenance of intracellular glutamate in Ha-ras-expressing cells. *Biochemical and biophysical research communications* 1995;211(3):878-884.

van Gunsteren, W.F., *et al.* Biomolecular simulation: the {GROMOS96} manual and user guide. 1996.

Vasudevan, S.V. and Balaji, P.V. Dynamics of ganglioside headgroup in lipid environment: molecular dynamics simulations of GM1 embedded in dodecylphosphocholine micelle. *The Journal of Physical Chemistry B* 2001;105(29):7033-7041.

Veluraja, K. and Rao, V. Theoretical studies on the conformation of β-DN-acetyl neuraminic acid (sialic acid). *Biochimica et Biophysica Acta (BBA)-General Subjects* 1980;630(3):442-446.

Venil, C. and Lakshmanaperumalsamy, P. Solid state fermentation for production of L Asparaginase in rice bran by Serratia marcescens SB08. *The Internet Journal of Microbiology* 2009;7(1).

Venil, C.K., *et al.* Production of L-asparaginase by Serratia marcescens SB08: Optimization by response surface methodology. *Iranian Journal of Biotechnology* 2009;7(1):10-18.

Verma, N., *et al.* L-asparaginase: a promising chemotherapeutic agent. *Critical reviews in biotechnology* 2007;27(1):45-62.

Wang, C.-C. and Lee, C.-M. Isolation of the Acrylamide Denitrifying Bacteria from a Wastewater Treatment System Manufactured with Polyacrylonitrile Fiber, *Current Microbiology* 2007; 55(4), 339-343.

Wei, D.z. and Liu, H. Promotion of L-asparaginase production by using n-dodecane. *Biotechnology Techniques* 1998;12(2):129-131.

Wriston, J. and Yellin, T. L-asparaginase: a review. *Adv Enzymol Relat Areas Mol Biol* 1973;39:185-248.

Wriston, J.C. 5 L-Asparaginase. *The enzymes* 1971;4:101-121.

Yaacob, M.A., *et al.* Characterisation and molecular dynamic simulations of J15 asparaginase from Photobacterium sp. strain J15. *Acta Biochimica Polonica* 2014;61(4):745-752.

Yamada, H. and Shimizu, S. Microbial and enzymatic processes for the production of biologically and chemically useful compounds [New Synthetic Methods (69)]. *Angewandte Chemie International Edition in English* 1988;27(5):622-642.

Yao, M., *et al.* Structure of the type I L-asparaginase from the hyperthermophilic archaeon Pyrococcus horikoshii at 2.16 Å resolution. *Acta Crystallographica Section D: Biological Crystallography* 2005;61(3):294-301.

Zhang, Z. and Friedrich, K. Artificial neural networks applied to polymer composites: a review. *Composites Science and technology* 2003;63(14):2029-2044.

Zhou, H. and Zhou, Y. Distance-scaled, finite ideal-gas reference state improves structure-derived potentials of mean force for structure selection and stability prediction. *Protein science* 2002;11(11):2714-2726.

Zöller, M.E., *et al.* Malignant and benign tumors in patients with neurofibromatosis type 1 in a defined Swedish population. *Cancer* 1997;79(11):2125-2131.

APPENDIX-A

Quantitative assay of L-asparaginase enzyme activity

Enzyme activity of the culture filtrates was determined at the end of cultivation time by quantifying ammonia formation in a spectrophotometric analysis using Nessler's Reagent. A 0.1 ml sample of culture filtrate (enzyme solution), 0.9 ml of Tris-Hcl buffer (pH 8.6) and 1 ml of 0.04 M L-asparagine solution were combined and incubated for 10 min at 37°C. The reaction was stopped by the addition of 0.5 ml of 15% (w/v) TCA. After centrifugation, I ml of supernatant was diluted to 3 ml with deionized water, treated with 1 ml of Nessler's reagent and 1 ml of 2M NaOH. The color reaction was allowed to proceed for 20 min before measuring the OD at 480 nm. The OD was then compared to a standard curve prepared from solutions of ammonium sulfate as the source. Blank was prepared by without asparaginase enzyme sample (produced in production medium).

Unit: One unit (IU) is defined as the amount of enzyme that released 1 μmole of ammonia from L-asparagine per minute at pH 8.5 at 37ºC.

Determination of μmole of ammonia liberated using the standard curve

Units/ml enzyme = [(μmole of ammonia liberated) (1) (2.5)]/ [(0.1) (20)]

Where,

0.1= Volume of filtrate

1.0= Volume of supernatant

2.5= Total volume of reaction mixture

20= Time of assay in minutes

APPENDIX-B

CONSTRUCTION OF AMMONIUM SULPHATE

STANDARD GRAPH

Standard graph was prepared by treating ammonium sulphate with trichloro acetic acid, NaOH and Nessler"s reagent.

Stock solution of 10 mM: It was prepared by dissolving 114 g of ammonium sulphate in 1000 ml distilled water in a volumetric flask. From this 1μM working standard was prepared by serial dilution.

Procedure:

1. From 1μM working standard solution, 0.4, 0.8, 1.2, 1.6, 2.0 and 2.4 μM standards were prepared by taking 0.4, 0.8, 1.2, 1.6, 2.0 and 2.4 ml respectively.

2. Total volume was made upto 2.5 ml using deionized water.

3. 1 ml of Tris-Hcl buffer was added to each test tube and incubated it for 30 minutes at 37°C.

4. 0.5 ml of 15% TCA was added followed by 1.0 ml Nessler's reagent and 1.0 ml 2M NaOH. Immediately the solution was mixed and allowed to react for 20 minutes.

5. Blank was prepared by adding 1 ml of water instead of ammonium sulphate solutions.

6. The final solutions were read at 480 nm in spectrophotometer.

7. A standard curve was constructed by taking ammonium sulphate (μM/ml) on X-axis and corresponding optical density on Y-axis.

S.No	0.05M Tris HCl(ml)	1 μm Ammonium Sulphate (ml)	Deionized Water (ml)	Incubate for 30 Min at 37°C	15% TCA (ml)	Nessler's Reagent (ml)	2N NaOH (ml)	Incubate for 20 Min at 20°C	OD at 480 nm
Blank	1.0	0.0	2.5		0.5	1.0	1.0		0
S1	1.0	0.4	2.1		0.5	1.0	1.0		0.242
S2	1.0	0.8	1.7		0.5	1.0	1.0		0.468
S3	1.0	1.2	1.3		0.5	1.0	1.0		0.699
S4	1.0	1.6	0.9		0.5	1.0	1.0		0.93
S5	1.0	2.0	0.5		0.5	1.0	1.0		1.153
S6	1.0	2.4	0.1		0.5	1.0	1.0		1.403

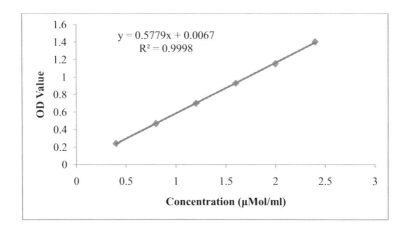

y = 0.5779x + 0.0067
R² = 0.9998

Ammonium sulfate standard plot

APPENDIX-C
Total Protein Estimation by Lowry's Method

Reagents Required

1. BSA stock solution (1mg/ml),

2. Analytical reagents:

(a) 50 ml of 2% sodium carbonate mixed with 50 ml of 0.1 N NaOH solution (0.4 gm in 100 ml distilled water.)

(b) 10 ml of 1.56% copper sulphate solution mixed with 10 ml of 2.37% sodium potassium tartarate solution. Prepare analytical reagents by mixing 2 ml of (b) with 100 ml of (a)

3. Folin - Ciocalteau reagent solution (1N) Dilute commercial reagent (2N) with an equal volume of water on the day of use (2 ml of commercial reagent + 2 ml distilled water)

Principle

The phenolic group of tyrosine and trytophan residues (amino acid) in a protein will produce a blue purple color complex, with maximum absorption in the region of 660 nm wavelength, with Folin-Ciocalteau reagent which consists of sodium tungstate molybdate and phosphate. Thus the intensity of color depends on the amount of these aromatic amino acids present and will thus vary for different proteins. Most proteins estimation techniques use Bovin Serum Albumin (BSA) universally as a standard protein, because of its low cost, high purity and ready availability. The method is sensitive down to about 10 μg/ml and is probably the most widely used protein assay despite its being only a relative method, subject to interference from Tris buffer, EDTA, nonionic and cationic detergents, carbohydrate, lipids and some salts. The incubation time is very critical for a reproducible assay. The reaction is also dependent on pH and a working range of pH 9 to 10.5 is essential.

173

Procedure

1. Different dilutions of BSA solutions are prepared by mixing stock BSA solution (1 mg/ ml) and water in the test tube as given in the table. The final volume in each of the test tubes is 5 ml. The BSA range is 0.05 to 1 mg/ ml.

2. From these different dilutions, pipette out 0.2 ml protein solution to different test tubes and add 2 ml of alkaline copper sulphate reagent (analytical reagent). Mix the solutions well.

3. This solution is incubated at room temperature for 10 mins.

4. Then add 0.2 ml of reagent Folin-Ciocalteau solution (reagent solutions) to each tube and incubate for30 min. Zero the colorimeter with blank and take the optical density (measure the absorbance) at 660nm.

5. Plot the absorbance against protein concentration to get a standard calibration curve.

6. Check the absorbance of unknown sample and determine the concentration of the unknown sample using the standard curve plotted above.

BSA (ml)	Water (ml)	Sample conc. (mg/ml)	Sample volume (ml)	Alk. $CuSO_4$ (ml)	Lowry reagent (ml)	O.D. 660 nm
0.25	4.75	0.05	0.2	2	0.2	0.050
0.5	4.5	0.1	0.2	2	0.2	0.086
1	4	0.2	0.2	2	0.2	0.111
2	3	0.4	0.2	2	0.2	0.226
3	2	0.6	0.2	2	0.2	0.319
4	1	0.8	0.2	2	0.2	0.400
5	0	1.0	0.2	2	0.2	0.490

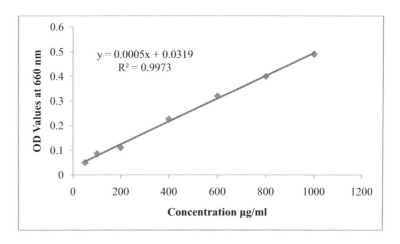

y = 0.0005x + 0.0319
R² = 0.9973

Protein standard graph

APPENDIX-D

ANOVA analysis of Percent Cell Viability as a function of Concentration of Enzyme ($IUml^{-1}$)

Social Science Statistics

One-Way ANOVA Calculator

Success!

Explanation of results

The output of this calculator is pretty straightforward. The values of F and p appear at the bottom of the page. If the text is blue, your result is significant; if it's red, it's not. The only thing that might catch you out is the way that we've rounded the data. The data you see in the tables below, which provide details about the calculation, has been rounded. However, we did not round when actually calculating the values of F and p. This means that if you try to calculate these values on the basis of the summary data provided here, you're likely going to end up with a slightly different – and less accurate – result.

Treatment 1	Treatment 2	Treatment 3	Treatment 4	Treatment 5
75.34	82.89	40.29	19.14	
74.99	92.79	46.82	14.97	
69.88	94.62	34.92	12.13	

Summary of Data

Treatments	1	2	3	4	5	Total
N	3	3	3	3		12
ΣX	220.24	270.17	115.83	43.24		557.48
Mean	73.4133	39.39	38.61	14.4133		46.4567
ΣX²	16157.8366	20598.7591	4895.0949	631.7374		31916.427
Std.Dev	3.1052	2.9361	3.5988	2.0621		23.3895

Result Details

Source	SS	df	MS	
Between-treatments	5946.8351	3	1982.2784	F = 223.57784
Within-treatments	70.9293	8	8.8662	
Total	6017.7645	11		

The F-ratio value is 223.57784. The p-value is < .00001. The result is significant at p < .05.

Calculate   Reset

APPENDIX-E

ANOVA analysis of Percent Cell Viability as a function of Time

A
Spoonful
of
Medicine

The Life of a GP Surgeon
By

Dr Keir MacKessack-Leitch
M.B., Ch.B., F.R.C.S.Ed.,
O.St.J.

2005

PUBLISHED IN GREAT BRITAIN BY
SANTIAGO PRESS
PO BOX 8808
BIRMINGHAM
B30 2LR
E-mail for orders and enquiries:
santiago@reilly19.freeserve.co.uk

Cover design by W.A.C.Dawson

ISBN 0-9539229-5-2

Printed and bound in Great Britain by
Antony Rowe Ltd., Chippenham, Wiltshire

The Author

~ P R E F A C E ~

It was Francis Bacon who wrote "Some books are to be tasted, others to be swallowed, and some few to be chewed and digested". This chronicle is simple fare, which I hope you might at least find palatable, if only as a sample of social history.

My aim is to give an account of my training as a doctor and a surgeon as well as sharing with you some of my varied experiences. This is mainly the story of my working life.

Today, such a hybrid as a GP Surgeon does not exist. Advances in medicine and technology are quite bewildering to an old practitioner like myself. It was my privilege to practice in earlier days when one was presented with a wide range of problems. So many of these fields are now the prerogative of the specialist surgeon. Specialisation is paramount, and rightly so. However, I am still grateful for the challenges I faced and would not change my training or the course of my chosen profession.

Keir MacKessack-Leitch
2005

~ A Spoonful of Medicine ~

The Life of a GP Surgeon

Chapter 1: A Farmer's Son

I was born on November 24th, 1919 in Grays hospital, Elgin, Morayshire. I came from a long line of farmers and our family home was at Carden Farm just two kms from the village of Alves and eight kms from Elgin. My mother was from Edinburgh and had aspirations of becoming an opera singer until she met and fell in love with my father.

We were very much in the country. Lighting at night was by candles and paraffin lamps, heating by wood and coal. Our water supply was provided by a windmill sited in a nearby field. Sometimes this tall windmill failed to work and we had to resort to using rain water collected in a large barrel near the kitchen door. You can imagine how cold that water felt in wintertime, especially since mother insisted that my brother and I had a cold bath every morning. In severe weather I can well remember our bath containing lots if ice .... Ouch!

What prompts a person to follow a career in a profession? Perhaps one of the parents, in particular the father, was a professional lawyer, a teacher, a scientist, a doctor, or a minister. Perhaps one has to have a specific "calling" for the last named. For myself, I was afflicted with attacks of asthma from an early age and, as a result, was often a very poorly child. The present day treatment of this most common and somewhat cruel disease is quite revolutionary compared with my treatment during the 1920s when I was a boy. Sometimes I was given special stramonium cigarettes to smoke and inhale. These gave me quick relief, but at the cost of making me light headed, dizzy and with a tendency to sweat. This was probably due to the lowering of my blood pressure. Later on, I used a wonderful brown flat-glass aerosol. The liquid inside this inhaler was called Brovon or Riddobron, made by a firm in Aberdeen. A tube with a rubber bulb at the end was attached and, by rapidly squeezing the bulb, a fine spray was released. Inhaling this gave me relief from asthma attacks for many years to come. I can well remember that our own family GP visited me quite often and was always kind. When I was twelve years old I had already decided I wanted to be a doctor and to help people get better. I loved reading anything and everything of scientific interest, and particularly with a medical bias. If I overheard an adult conversation discussing the illness of a relative or friend, I would retire to my father's study and look up the definitions of the mentioned diseases in the Chambers Encyclopaedia.

It is likely that I may have inherited a leaning towards medicine from my mother, herself a most keen first-aider. She seemed to revel in treating not only her family but also anyone who sustained injuries on our family farm - human and animal alike! My father owned a traction engine which even sported a roof - very smart for the 1930s. The flywheel was used to drive the threshing machine at harvest time. One day, the large, broad strap running from the flywheel to the harvester caught one of the workers on his head, nearly scalping him. My mother, of course, was summoned. She quickly appeared at the scene, armed with a large box of her "essential" first-aid supplies, and dressed his wound. I helped her. There was a lot of blood about but this did not bother me. The victim, who was considerably shocked, was taken off to hospital post haste. He returned to work the following week.

My younger brother, David, and I were both prone to falling off our bicycles on a very regular basis, with the usual resultant knee abrasions. These were immediately treated enthusiastically by our mother with hot bread poultices and, sometimes, steaming porridge! This of course was repeated daily and resulted in septic wounds. However, these wounds were eventually resolved by the judicious use of liberal applications of iodine. Boy, did that stuff sting!

David and I were given pocket money - one penny a week only, although sometimes my grandmother supplemented this. We spent our money on liquorish straps or sherbets or even oogo-pogo eyes. The latter were boiled sweets, you could buy eight for one penny, and they fascinated us by changing colour no less than four times as they were sucked! During school holidays, I spent much of my time working on the farm. I even had my own horse, called "Sparrow". She was a Clydesdale and immensely strong. I could not groom her at the end of the day for I was allergic to horse dandruff which immediately gave me asthma. I was fortunate that asthma did not curtail any of my other activities because I loved sport of all kinds. I was an addict to physical fitness. Work on the farm was sometimes very arduous and demanding - there were few mechanised implements in those days. In the winter holidays, I lifted turnips, topped and tailed them and loaded them onto a cart. It was cold and hard work and my hands became so tough that I could take hold of a bunch of stinging nettles without being stung. Little did I know how precious these hands would be to me when I became a surgeon!

Another extremely hard job was the emptying, with other farm workers, of the byres and folds of dung and loading it on to carts. These folds, which had held cattle over the winter, were four feet deep with manure of straw and cattle dung. This had to be spread by fork over the fields before ploughing. Very little artificial manure was used, as it was expensive, so organic farming was fully practised. Later, of course, these tasks were taken over by tractors, hydraulic lifts and mechanical muck-spreaders.

Summer holidays meant harvest time - getting up at 5a.m. and finishing around 7p.m. I became an expert at loading the cart with sheaves. I tried to build up to eight rungs high, the usual being six. Sparrow pulled this heavy load only if I led her with long reins and standing far back. If I held her by the head she refused to budge an inch!

How times have changed since I was a lad! My father farmed arable land of 650 acres. His livestock consisted of Aberdeen Angus and Hereford cattle. He was a leading breeder of the Large White pig. He also had a wide-ranging stock of poultry. He employed 15 men and 2 dairywomen. The latter milked the cows by hand and looked after the dairy which supplied all staff and workers with milk and butter. Three small horses were kept for riding or for helping with light work but the 16 Clydesdale horses were all eventually replaced by tractors. One tractor could do the work of six horses. I am glad I grew up when I did and can look back with such fond memories on an earlier and more romantic age. And of course, the lifestyle proved to be a wonderful training ground for the rigours of my future profession. The hard physical work conditioned me bodily and mentally for my life as a doctor and surgeon.

As a family we had a fortnight's holiday once a year at the end of June and beginning of July. This was the best time of year for my father to get away from farm work. The hay had been cut and stacked, the silage was soon ready to be lifted and the cereal harvest of oats, barley and wheat would not be ready until August. Like all farmers of the time in the 1930s, father was very cash strapped, so we went camping. For five to six years we always motored to the North West of Sutherlandshire usually choosing an ideal site with breath-taking views at Strathan Bay, six miles from Scourie. The journey was hazardous, for the roads in that area were narrow and rough, with grass growing in the middle. Our car, which was a Wolseley two seater with a dickey, could only get half-way up one steep hill. We all had to get out and let father reverse to the bottom and then accelerate at full throttle so he could drive it to the top. These were very happy days together. My parents had a large 6-foot high ridge tent whilst my brother and I slept in a very low tent into which we had to crawl. It was simply a dark green thick canvas tarpaulin - the edges expertly sewn by mother to take hand-made pegs. The tent ridge and poles were simply fashioned by father from larch tree branches. When it rained, the tent leaked here and there, but by placing one's finger over the raindrop and drawing it down the side of the tent to the ground, the rainwater simply ran down the tent following the finger track! Our first task was to cut masses of bracken with a sickle and load this into two large sugar beet sacks. These were our mattresses. Mother and father had canvas camp beds. There was a stream running nearby where David and I fished for trout using worm as bait. We were often awakened in the morning by the plaintive call of buzzards, which were nesting on a high crag rising vertically from the opposite side of the bay. Other bird life was fascinating, especially flocks of solan geese or gannets. These birds, with a wingspan of 1.5 metres, would fly 25 metres high and suddenly fold their wings and plunge dive into the sea to catch a fish. They rarely missed. We also occasionally saw a whale. A large seal often visited our bay and gazed at us. We called him "Jehosephat" We all went bathing daily either in the stream, which we had dammed to make a large pool, or in the sea. The sea was sometimes warm - perhaps thanks to the Gulf Stream. One day, when the tide was high and the bay full, there was an invasion of a shoal of young herrings. This was quite extraordinary, for the bay was so full of fish one could not even see through to the sea bottom.

We had to collect our milk from a sheep farm three miles away. Living there was a delightful family of two brothers, a sister and an elderly mother. The father had died some years before. They had 3,000 acres of hillside and to see their three collie dogs collecting their sheep was a wonderful sight. It was great fun to help them at the sheep-shearing, all done by hand. Of course, other neighbours helped as well. One year a local well-known farmer jumped into the loch to retrieve a sheep. Being a traditionalist, he had worn his kilt without much underneath! Unfortunately, the shock of the cold water gave him a fatal heart attack. In his will he left three dozen bottles of malt whisky, probably Macallan or Glenfarclas, all to be consumed at his funeral.

On Sundays, we could not collect our milk for their house was firmly shut and all blinds were drawn. We motored to Scourie to shop. There were about 30 houses scattered over a wide area and only one shop, which was a general store. There was an adjoining hotel of four bedrooms. One day, we took a boat trip to the wonderful bird sanctuary of Handa, famous for having the largest colony of guillemots in Britain. We arranged to meet the boatman at 11a.m. but he did not show up until 2.30 p.m. Time meant nothing in the North West of Scotland! We landed at the only beach on the island and walked uphill all the way to the other side. As we approached, the sound of seabirds was deafening and the smell of guano was quite awful. The whole of this side of the island ended in sheer cliffs 100 metres high (328 feet). Seabirds of all kinds were nesting on the cliff edges and I shall always remember the Great and the Arctic Skuas swooping alarmingly low over our heads. There was a pillar of rock called The Stack standing some 200 metres away and the same height as the cliffs. No birds had nested on the top of this stack since many years ago it was invaded by men who took all the eggs and young birds from there. The last inhabitants of Handa Island left in 1847 when their potato crops failed. The memories of our family camping days at Strathan Bay are very dear to me. Luckily, we did not know at the time that one day this lovely small bay was to become the site of a huge salmon farm. Our campsite would be bulldozed to make way for a road and a slipway to the edge of the sea.

I was educated at the Elgin Academy in the North of Scotland, an excellent co-educational school. Only seven of us made up the 6th form – six boys and one girl. The girl graduated MA at Aberdeen University and later returned to our school as a teacher. All six boys went to university to study medicine - three to Aberdeen and three, including myself, to Edinburgh. And we all became doctors – I am sure this must be a very unusual coincidence.

We did not have an OTC at our school but instead had a fortnight's mountaineering expedition to the Cairngorms in the Grampian Mountains. Our camp was near Aviemore, close to a sandy beach on the shore of Loch Morlich. The summit of Cairngorm was visible from the camp, 1,245 metres high. Any mountain over 1,000 metres, i.e. approx 3,000 feet high, is known as a Munro. The camp consisted of five bell tents, each accommodating up to eight boys. We were all 15 to 17 years old and four or five masters accompanied us. The masters slept in tents of the ridge type. We slept on the ground and that was hard! Some boys had substantial sleeping bags, but mine was just a sheet and two blankets stitched together by my mother. One of our first tasks was to build a large trench to serve as a latrine. We all took it in turns to cook, peel potatoes and prepare meals. One day involved climbing, followed by a day in camp. This was not really a rest day for we were encouraged to go swimming, sailing or canoeing on Loch Morlich. At all times, we were chaperoned by our teachers, and, with just a few exceptions, we boys thoroughly enjoyed our camping holiday. If the occasional lad was unhappy and unable to cope, he was sent home.

The wildlife was a great source of interest. Lots of ptarmigan were seen, especially at 1,000 metres up on the slopes of Cairngorm. They were difficult to find for their camouflage was so good. On approaching a ptarmigan, it would freeze and not fly away, trusting its colouring to make it invisible. These birds were all white in winter. An occasional dotterel was seen - a rather stupid bird with a reddish breast which also remained still when approached. It was hard to believe that this dotterel had migrated all the way from North Africa, where it had spent the winter before returning to breed in the high altitude of the Grampians. One was very occasionally lucky enough to see the magnificent golden eagle with its wingspan of over two metres and standing one metre high. And sometimes a herd of red deer was spotted.

In my final year at school and on my final camping expedition, two boys and myself, together with one of the masters, decided to tackle as many "Munros" in one day as we could. We camped overnight near the base of Cairngorm in Corrie on Trecht, close to a small loch. We all had proper sleeping bags this time - I was given a super one filled with eiderdown. We woke up early the next morning and first of all bathed in the loch. Gosh! That water was cold, for it had tumbled down from snowfields. We then set off and climbed Ben Macdui (1,309 metres) approaching the summit on the "dairy-maids fingers" - huge finger-like lanes of snow. Large tracks of snow remain in the Cairngorms all the year round, probably the largest being the White Lady to the side of the Cairngorm mountain. After reaching the top of Ben Macdui, we then decided to cross the Lairig Pass. A huge and steep snowfield lay in our path. Our teacher, who was a junior mathematics master known as Pluto, simply ran across this snow digging his climbing boots in as he went and yelling at us to do the same. We three boys were only wearing heavy walking shoes and we gingerly climbed onto the snow slope, the other two in front of me. Suddenly, one boy lost his footing and slid at breakneck speed down the slope where fortunately Pluto managed to catch him. They both fell in a heap. The other boy started to come back towards me but he also slipped and fell and unfortunately somersaulted at the bottom onto the rocks. Pluto managed to catch the boy's haversack, which probably saved a serious accident. It was all very frightening. I managed to very carefully climb off the snow and onto firm land and then took a very long time to climb down round the snow area. Pluto should have taken this route in the first place! My two friends were both somewhat shocked, but after some hot sweet tea from a thermos flask, some chocolate and raisins, we all agreed to continue our quest. Well, we crossed the River Dee, just a small stream there and running the length of the Lairig Pass. We next climbed Devil's Point (1,004 metres) and then Cairn Toul (1,219 metres) and finally Braeriach (1,296 metres). As we approached the top of each mountain, we all lifted a stone and placed it on the Cairn, which marked the summit. This was a custom when climbing mountains. We rested for a bit on Braeriach but quite suddenly saw mist appearing. We all knew how to read a compass and quickly took a bearing for our route back to the Lairig Gru. It was essential to do this, as huge cliffs which marked the source of the River Dee were dangerously close to our right. Slowly, we descended, talking often to each other because visibility was down to 10 metres. As we climbed eventually up the Ben Macdui side, close to the so-called Lurchers crag and following the Marsh Burn, the mist was beginning to clear. When we reached the top, suddenly we could just glimpse our small campsite! My two companions were completely exhausted and went straight back to camp. But Pluto and I decided to go on and tackle Cairngorm. So we succeeded in climbing **five** Munros in one day, which was quite a feat. We reached camp late in the evening and were congratulated all round. One can imagine that our sleep that night was pretty sound! The headmaster later reprimanded Pluto for his foolhardiness.

To gain acceptance as a medical student at Edinburgh University, I had to attend for an interview, have a written recommendation from my school headmaster, and had to pass the necessary categories of A levels or "Highers" as they were called in Scotland. I attained the required Higher grades in English Language and Literature, Mathematics, Chemistry and Physics. My father accompanied me to my interview which was with the Dean of the Faculty, Professor Sir Sidney Smith, Professor of Forensic Medicine, a subject which became one of my favourites in the years to come.

Chapter 2: Medical School

It was a truly wonderful yet daunting experience to be accepted and to commence my career at Edinburgh University in September 1937 at the age of 17 years. There were 230 medical students in my year, 25 of whom were female. Two-thirds of the students were from Scotland, the remainder mostly from England and Wales with a small number from distant parts of The Empire, as it was then. Most of these would eventually return to their own countries. In time, 36 of those from Britain emigrated and embarked on medical careers abroad.

Fortunately, I was in a hostel - Cowan House in George Square - with about 80 other students, all from different faculties. Cowan House and adjoining buildings were replaced many years later by a magnificent library - one of the largest in Europe. About one third of us were first year students or "freshers", as we were named. We were allowed to stay in the hostel for 3 years and then had to face the problem of finding digs for ourselves. Shortly after my arrival, I learnt, with some trepidation, that there was an initiation ceremony for the freshers. Each new student had to perform something whilst standing on top of a table, for example, recite poetry or sing a song. We were all herded outside the large sitting-room and called in by name, one at a time. Upon entering, the room was in complete darkness. I was pushed towards the direction of the table, and gingerly climbed up onto it. Once on the table, a bright light was positioned so that it shone directly into my face and I was told to begin. Having a deep baritone voice I sang a song by

Paul Robeson called "Lonely Road". I felt more than a little nervous and was showered with rotten tomatoes and the odd egg. But this was not all, for the table was tipped dangerously from side to side amongst many derogatory cheers, and, of course, I soon fell off! My initiation over, the next victim was ushered in.

My room in the hostel was off the ground floor corridor, known as "The Slums". One evening, through unusual circumstances, I acquired some precious bones. It was quite dark outside and I was busy studying. A terrific argument started in one of the rooms above me - not an unusual occurrence - but this particular evening, it seemed more serious than normal. I heard a huge thump outside my window, which overlooked a red all-weather tennis court. This was rapidly followed by more thumps. I quickly turned off my desk light and peered out of the window. Lo and behold, I could just make out several human bones lying scattered across the tennis court! The noise abated upstairs. No one came down to collect these valuable objects. I kept an eye on them and, after two days, as they were still unclaimed, I nipped out of my window late at night and retrieved them like a prized booty! I did not inform anyone of my "treasures" and no one reported their loss.

By chance, soon after, I managed to buy a box containing a few more human bones from a senior student who was a bit strapped for cash. I added these to my "tennis court" collection and when assembled they almost made a half-set. This was invaluable to me in my study of osteology. In later years, when I was a practising doctor, I often made use of my bone collection during lectures to students, nurses and first-aiders. Once, I even sold them to the Nursing School of my local hospital. Later, the School acquired a new skeleton so the Senior Tutor kindly returned my bones free of charge! Now an elderly retired doctor, I still have my collection in a box, somewhere in the attic.

One of the first things I did at Edinburgh University was to purchase various text books that medical students were advised to have. I managed to buy all mine second-hand and in good condition. I was lucky. When I looked at some of the books, in particular the anatomy-dissecting manual, I was horrified to read the huge long words that seemed like a frightening foreign language. At one stage, I felt I just could not cope with remembering these words, let alone pronounce them! But after talking to my fellow students, I found they were in exactly the same frame of mind and this encouraged me to continue.

My first introduction as a medical student to anatomy was in the Anatomy Department. We were divided into groups. Each group was given an appointment with the senior anatomy lecturer, Dr E.B. Jamieson. We had to congregate outside his private study until called in. His book, which could be readily pocketed, was the "bible" of anatomy and known as the "wee Jimmy". When it was my group's turn to be ushered in to see the great man, I noticed that many of those coming out looked shaken and very pale. Well, when I went in, there was "Jimmy", wearing a black skull cap. He had piercing blue eyes and a determined, overpowering facial expression. He told me to sit down and said, "Your name?" He asked a few more questions and then, to my amazement, I saw on his desk a whole head and neck cut in two, showing the brain, throat, and blood vessels of the neck. I had never seen a dead body before, let alone a severed head! But this did not particularly upset me. Then I noticed one of my fellow students stretched out on the floor behind the desk. I asked Dr Jamieson if I could help. He simply said, "No, he's fainted, leave him. Good morning." I replied, "Good morning, Sir", and left. This experience crystallised my ambition to study medicine and to become a doctor. I became addicted to anatomy - it was one of my favourite subjects.

Anatomy is such a positive subject to study. I eventually gained honours and a first class certificate. I learnt how beautifully and wonderfully made we all are. I never cease to admire how our Creator fashioned us from a single fertilised cell to the full development of Homo Sapiens. From the scientific point of view, this is all just the embodiment of evolution. I am sometimes reminded of a quotation which sums up my feelings: "In studying medicine, one wonders at the engineering marvels of the human body. One becomes more and more impressed by the wisdom of the Grand Creator. One is moved to say to God, as did the Bible Psalmist of long ago – 'I shall laud you because, in a fear-inspiring way, I am wonderfully made.'"

My five years as a medical student were, without doubt, very happy and I have many wonderful memories. Anatomy, the study of the structure of the human body, was part of the first year's curriculum, coupled with physiology, the study of the normal functional and processes of the human body. I will never forget going into the Anatomy Department armed with my small box of dissecting instruments and manual and wearing a long, light brown overall. This was it! I was about to perform my first dissection! The room was huge and at the top of the medical buildings. Furthermore, it was lit with many skylight windows and overhead lights, the latter being on pulleys so that they could be pulled down over a body as required. There must have been at least twelve corpses, each laid on a large stainless steel table. Three students were attached to each body part, which was dissected and studied for a whole term. My trio started with the leg. Neither of my two companions wanted to be the first to cut the skin of the thigh and they voted me to do so. It was indeed a challenging moment.

It was difficult to cut through the thick skin of the corpse, well preserved in formaldehyde. My hand did not shake. There was no bleeding! We had to dissect and identify all of the structures under the skin or epidermis, parts such as fat, fascia, muscles, nerves and blood vessels. The inter-relationship of one structure to another defines the study of anatomy. And so our careers commenced.

Work was hard and it was necessary to study for many hours each day, often until midnight. We had to read many books and learn hundreds of completely new words (and spell them correctly). Gradually, our vocabulary increased and discussions between students enhanced our knowledge of medicine. When medical students talk about their work and studies, their language would be quite unintelligible to any lay person. We attended many lectures from 9a.m. until 5p.m., interspersed with demonstrations and experiments. Countless exams had to be taken, all of which had to be passed. If you failed twice, you were in acute danger of being asked to leave, and then probably to be called up to serve in one of HM Services. The Second World War had begun.

I very seldom went out in the evening, but reviewed my day's work and read my books. I would work for an hour or so then wash my face in cold water, and get down to my books once again. Sometimes I listened to music - mostly light classical - either on my small wireless, a "Tukaway", or on my gramophone, a huge box affair with a large megaphone and supplied with five large "78" records. My favourites were "Meditation", "Invitation to a Dance", "William Tell", Chopin's piano recitals, and some Scottish dance tunes.

It took me several months to form close friendships with the other students - some medics and some from other faculties. We became typically anti-establishment, railing against the ruling party and against authority in general, changing our allegiances with surprising ease! It is difficult now to picture myself wearing a bootlace instead of a tie - my way of protesting against all things conventional! Of course, as we matured, we grew out of these extremes - but it was great fun at the time and very liberating.

My pocket money was initially £1 a month and anything extra that I saved from my very meagre wages earned on the family farm. By this time I had started smoking cigarettes which I limited to 6 or 8 a day. My choice varied between Craven A, Players Weights and Woodbines.

Saturday was a day of sport for me. I played rugby in the autumn and spring terms, and tennis in the summer. But I also took part in other sports - hockey, lacrosse and badminton and I even once boxed for my university! I lost on points in the latter for my opponent was a professional boxer! My nose was well bloodied at the end but his was too – a little bit!

Saturday evening was the highlight of the week. I always went to the Men's Union dance in Teviot Place with a few of my pals but first visited the odd pub, our favourite being "The Hole in the Wall" where we all drank beer, usually McEwans, but occasionally an extra strong drink called "Tiger Juice". I loved dancing – ballroom dancing – and if I found a good partner, usually from the domestic science college, I stuck with her and chaperoned her home at the end of the evening. This frequently meant a long walk to Atholl Crescent, where the college was, and I usually took my bicycle with me. If I happened to take a girl to the cinema one evening, and she did not pay for herself, I could not afford to ask her again!.

Despite my rebellious tendencies, I still went to church – St John's in Princes Street at the west end side of Edinburgh - on most Sundays. I always wore my kilt and after the service enjoyed walking through Princes Street gardens, which were so lovely in the spring and summer. The flower clock at the end of the gardens was a gardening masterpiece and I often waited there till the hour came up and the cuckoo called the time of day. Sometimes I walked on to Waverley Station and then up the 180 or so steps to Princes Street itself. One had to be very wary of walking up those steps for the draught of air was strong enough to blow one's kilt to an embarrassing height! Women also had to hang on to their skirts. During the summer, on Saturdays, I quite often went ice-skating at the rink in Haymarket and even learnt to waltz on the ice – a little! I visited all the famous sites in Edinburgh, such as the zoo, the British Museum, Sir Walter Scott's monument (in Princes Street gardens) – what a lovely view from the top of it. And on 1st May, I even climbed Arthur's Seat at dawn to wash my face in the dew!

Many of my teachers and lecturers were famous men and some of world renown. Dr E.B. Jamieson (Jimmy), our lecturer on anatomy, had the most remarkable memory. Some students said he remembered the names of their respective fathers. He gave us an hour's lecture, on most mornings of the week, at a slow pace. I was able to write down what he said and some students even bought "Jimmy's notes" from the medical bookseller. His lectures were word-for-word perfect from one year to another – an amazing feat.

Professor Brash, the professor of anatomy who lectured on the science of anatomy, often digressed to evolution and anthropology. He had a habit of saying, "It's a very remarkable fact." Whenever he repeated this, the whole class (all 200 of us) stamped our feet or thumped our desks with a book – just once – an almighty crash! It was Professor Brash who was partly responsible for accusing Dr Buck Ruxton for the murder of his wife. Dr Ruxton had dismembered his wife's body and distributed parts of it in the countryside. A leg was found where the hip was separated from the pelvis, but the ligament connecting the head of the femur with the acetabulum of the pelvic bone was cleanly cut through. Only a doctor would have known about this ligament.

Another of my anatomy teachers was Dr Walmsley who later became Professor of Anatomy at Dundee University. He was the most wonderful artist. He had the ability to draw in detail with coloured chalk on the lecture board, illustrating the relation of one anatomical structure to another, whilst lecturing at the same time!

In the third year, we progressed to the Royal Infirmary. We learnt to converse with patients and to examine them with our hands and our instruments and stethoscopes. I was fortunate in being appointed Ward Clerk to Professor Sir Derrick Dunlop's ward. He held the Chair of Therapeutics and Clinical Medicine. He was the embodiment of the British gentleman. He was a tall, dignified and courteous man. All applauded his lectures. He once wrote, "Just as the old horse and buggy, though very slow, causes near fatal accidents, whereas the modern motor-car is a lethal instrument, so the old fashioned bottle of medicine, though relatively ineffective, was also comparatively innocuous, whereas the modern medicine, like atomic energy, is powerful for evil as well as for good." I realised that the problems presented on the wards were the best instructors in the art of medicine and in the understanding of human nature. We students were propelled into a steep learning curve at this point of our training.

But it was the surgical wards that fascinated me. One evening, I was invited to watch the removal of an appendix by a senior consultant surgeon who was a distant relation of my family. I sat alone in the gallery, consisting of several tiered marble seats surrounding one half of the operating theatre, looking down onto the operating table. I am ashamed to say that I did not witness the whole of the operation and had to leave! The smell of ether was overpowering and when the surgeon made the first incision and blood appeared, I felt a sensation of pain where I had had my own appendix removed when I was a boy. Luckily I did not have to withdraw from any further operations! The surgeon's skill fascinated me, as did a pathologist I watched whilst performing a post-mortem and lecturing to students simultaneously, explaining the procedure and investigation involved. I wondered if perchance I might attain the necessary skill and knowledge to do the same one day. I knew from that experience that I would become a doctor. This reminds me of the quotation from Trousseau, a lecturer on Clinical Medicine, and who wrote this tribute: "Literature, painting and music do not yield an enjoyment more keen than that which is afforded by the study of medicine, and whoever does not find in it, from the commencement of his career, an almost irresistible attraction, ought to renounce the intention of following our profession".

I was soon to learn that the study of medicine is of eternal and absorbing interest. I supplemented my student education in lectures and ward work by frequently going to watch post-mortems, attending medical and surgical out-patient sessions and going to the Accident and Emergency Department at the Royal Infirmary. It was here that I learnt to dress wounds, many of which were grossly infected and stank unbearably. I was also instructed in the art of stitching up clean wounds; a never-ending stream of casualties kept coming in for treatment. This was all a wonderful experience for me. My hard work was rewarded when I attained first class honours in my pathology exams and was also awarded the first prize in clinical surgery.

In the realm of therapeutics and pharmacology, medical students were taught how to prescribe and to dispense medicines. The apothecary dosage system was used - grains, mins, drachms and ounces. When writing a prescription we headed it with the sign of Jupiter - Rx. Any newly discovered remedy, such as the antibiotics, were labelled in the metric system. Later in my career, when the metric system was fully adopted, it was quite a challenge to change dosages from grs. to gms. and mins. to mls. or ccs.

In my fourth year, I applied to become a resident at the Cowgate Dispensary, along with a great friend of mine, Reginald Bosanquet, an Australian. We were there for six months. It was essentially a preliminary training post for medical students aspiring to become medical missionaries. My friend and I had no ambition for this calling, but the experience we gained in working there was immense. The so-called Chief was a graduate doctor and he was soon to become a missionary in China. There were about nine of us and we took it in turns at weekends and at night-time to be on-call. My doctor's bag was a small Gladstone bag that had once belonged to my grandfather! I normally used it to carry my rugby boots and kit. The doctor in charge gave each of us on-call the necessary equipment for emergencies. I carried my own stethoscope, auroscope (for the ears), ophthalmoscope (for the eyes) and a second hand sphygmomanometer. The latter was used to record the blood pressure. On Sunday afternoons, I would admire my resident companions who wanted to become missionaries as one or two of them would stand on a make-shift platform and preach the Gospel. Quite often, a congregation would collect of between 20 and 30 people, all from this slum area – the Grassmarket, Cowgate, Canongate and High Street.

It was indeed a tough area to live in, especially in the evenings when drunken brawls were commonplace. My bedroom window – two storeys up – looked over a street, and sometimes I would hear women using dreadful language, shouting and screaming invective. They were often "high" on Brasso or methylated spirits which could cause blindness if taken in excess. One evening, I was called to the Men's Hostel in the Grassmarket to deal with a very drunk man who had a large wound to his scalp. There was blood everywhere – he had been hit on the head with a bottle which had smashed. It really was quite a picnic, as other inmates looked on, most of them also pretty inebriated. I quickly washed my hands with carbolic soap and proceeded to clean and then stitch up this huge wound which required at least eight stitches. There was no need to inject a local anaesthetic for the systemic effect of large quantities of alcohol was sufficient to neutralise any pain! In any case, the victim involved was singing at the top of his voice all the time whilst being held still by a few of his mates. There was heaps of laughter and a lot of swearing and one of his mates fainted, which added to the overall excitement. Stitches successfully in place, I finally doused the whole area with iodine and gave the patient an injection of tetanus toxoid. The next day, I called at the hostel and all was well. Seven days later, I removed the eight or so stitches and the patient was most grateful and really quite pleasant.

It was rather humbling sometimes when I was welcomed into a slum flat. Late one evening, I had to climb up four storeys of stone steps, very badly lit and with windows here and there which had no glass in them. It was the time of the "black-out", so I just had my pen torch with me. At the bottom of the stairs, a policeman told me that he would certainly not venture up by himself. But this did not deter me – I was the doctor with his "wee black bag". On another night call, I had to attend a woman in a high tenement house. Her flat was at the top. It was pouring with rain and water was leaking from her ceiling in two places into strategically placed buckets. This young good-looking woman was in a large wooden bed with rather grubby blankets. Three small children sat at the bottom of the bed. The patient had acute lower abdominal pains, which had come on quite suddenly when she was sneezing. It did not appear to be an acute appendicitis. I admitted her into the Royal Infirmary as an emergency. I later heard that she had ruptured an artery in the lower right abdominal wall – the inferior epigastric artery – an unusual accident. She made a full recovery.

It was very difficult to cycle around Edinburgh at night as the "black-out" during the war years was extraordinarily effective. In the colder months, smoke belched out from the many chimneys. This, combined with the frequent overhanging mist and fog, must have made the city well nigh invisible from the air. Small wonder that it was known locally as "Auld Reekie".

On a number of occasions, I had to attend an elderly man with pneumonia. He was very ill and he flatly refused to go into hospital. Fortunately, he responded to sulphonamide therapy and, when he was better, he was very grateful to me and offered me his bag-pipes which were in a rather dilapidated case. I declined his kind offer as he still had a rather nasty productive cough with slightly yellow sputum. I couldn't play the bag-pipes anyway, I only played the chanter, and that not at all well!

I recall one sad case of a two year old little child from a family of seven children, who lived in a very overcrowded flat. The child, a boy, had acute tonsillitis. After two days of treatment with sulphonamide, there was no improvement. I called in the doctor in charge of the Dispensary and he was suspicious of diphtheria but advised me to wait another 24 hours. The next day, I found the child no better and now very ill indeed. I sent him immediately to the infectious diseases hospital with a diagnosis of acute tonsillitis and possibly diphtheria. I heard later that a throat swab was taken which confirmed the diagnosis - diphtheria. The child was immediately given diphtheria antitoxin injection, but unfortunately to no avail for the little lad died 48 hours later of heart failure. I visited his family afterwards, but it was accepted that diphtheria, in those days, was, sadly, so often fatal.

My experiences as a medical student GP at the Cowgate Dispensary sowed the seeds of my desire to become a General Practitioner.

In my final and fifth year, I studied obstetrics and gynaecology. In midwifery, we had to be responsible for delivering a minimum of twelve babies. Part of the time we had to stay in a hostel within the precinct of the Simpson Maternity Hospital. Some of the medical students elected to do their midwifery elsewhere, such as at the Rotunda Hospital in Dublin. Each one of the cases we were responsible for had to be fully written up. Our book of delivery cases had to be handed in to our lecturers and marked accordingly. We, of course, had to witness a number of normal deliveries, and also problem cases when obstetric forceps were required. I was fortunate in being shown how to use these forceps. I remember the first baby I delivered was 9lbs 2 ozs. This was the first baby of a 16-year-old girl and all went well, without a hitch. Each medical student had to attend a certain number of home confinements in the company of a nurse – a district mid-wife. I had to live in Leith and my digs were not very far from the nurse's flat. I had my bicycle with me. It was a long way for me to cycle to the university every morning for lectures, but much quicker than walking. My cases were generally at night and, happily, the mid-wife I worked with was very professional and taught me a lot. I even delivered a breech and all of my deliveries, fortunately, went without complications. I did catch more than the odd flea, though, for they abounded in nearly every household!

Finally, in July 1942, the great day had arrived. After five years of study, I was "capped" at the McEwan Hall in Teviot Place, Edinburgh. I was one of about 140 students who graduated out of the original number of around 230. The very next day, I registered with the Medical and Dental

Defence Union of Scotland which cost me £3 – in years to come that sum was to become more than £1,000.

Chapter 3: Early Years as a Doctor

I was now a doctor. And with M.B.Ch.B. after my name. I thought I knew everything! This attitude was quickly dispelled when I started my first job. It is said that each one of us, however old, is still an undergraduate in the school of experience. I became a House Surgeon at the Western General Hospital under Professor Learmouth who later was knighted and became Professor Sir James Learmouth. He had operated twice on King George VI, once to improve the circulation of blood to his legs – a lumbar sympathectomy - and the second time to remove part of his lung – a lobectomy, for the King had a cancer of the lung. We generally knew the professor as "Pop". The highlight of his routine was his visit to the two surgical wards, which occurred two to three times a week. His entourage consisted of up to twenty visiting medical men and women and students and nursing staff. He walked in front of everyone along the corridors, and at a fast pace. In the wards, his House Surgeon had to know the name, age, address, diagnosis and case history of every patient in both male and female wards, each ward having 30 patients. It was quite a daunting memory task.

I received my calling up papers after only three months at the Western General. I was naturally disappointed to leave but excited about joining the army as a 1st Lieutenant in the Royal Scots. I had already done all the necessary square bashing and drills – at the Redford Barracks – and was even made a Corporal whilst a medical student. And so I was fully kitted up to go to North Africa, and had some instructions in warfare and survival and marching experiences along with some insight into front-line surgery, casualties and maintenance of a first aid unit. However, I developed an attack of asthma during a firearms exercise and required treatment. Following this, I underwent a medical by an army specialist and he said I was unfit for further service in the armed forces. I was thus discharged. This was a shattering blow for me. My brother, David, whom I dearly loved and admired, in turn received his call up papers. He became an officer in the Royal Corps of Signals. He served in the 7th Armoured Division under General Montgomery in North Africa and became a Desert Rat.

After a few days at home, which gave me time to digest my disappointment, I telephoned the Western General and explained to the secretary of my Chief, Sir James Learmouth, that I had been discharged from the army. I was duly summoned to his private office in the medical school. He invited me back as his House Surgeon to complete my six months' term. I was delighted to accept. During this time the sister-in-charge was so helpful to me and taught me the elements of general nursing care, such as the art of lifting patients, adjusting pillows and even bed making. Here I also learned about the dressing of wounds, often a daily task as so many were infected in those days.

One of my fellow doctors pricked his index finger when opening an abscess. He rapidly developed a severely infected finger and, later, septicaemia. He was treated with sulphonamides. Professor Learmouth had to operate on him and, at one stage, amputation of his arm was seriously considered. Very fortunately, the antibiotic therapy stopped the spread of infection. Amputation of the index finger only was sufficient surgery and he was able to continue his career as a doctor.

Near the Western General was a hospital for the mentally sick and insane. This was converted to a medical and surgical hospital and became the first Polish hospital in Britain, called the Paderewski Hospital. I became friendly with some of the Polish doctors and nurses and thereby learnt a smattering of their language. This helped a great deal when I had to treat Polish war casualties.

When my term of six months came to an end, I had an interview with Professor Learmouth. He asked me what I had in mind for the future. I told him I was interested in surgery. To my surprise and delight he proceeded to offer me two posts, the first as an assistant with a view to a partnership with a colleague of his – a doctor in Fort William. This held out a good future. The second was a position as House Surgeon to Mr Sangster, Consultant Surgeon at Ballochmyle Emergency Medical Services (E.M.S.) Hospital in Ayrshire. I decided upon the latter post and Mr Sangster appointed me when I went for an interview. And so I moved to Ayrshire, complete with bicycle! I was very happy there and enjoyed working with Mr Sangster, F.R.C.S (of both London and Edinburgh).

We were alerted one evening to receive casualties. A Merchant Navy cargo ship had been torpedoed and sunk in the Firth of Clyde by a German U-boat. I do not think this ever reached the ears of the National Press. We admitted about 15 casualties, mostly suffering from burns caused by sailors jumping from their sinking ship into the sea where surface oil was on fire. I remember one poor man who was burnt over most of his body. It was the first severe casualty I had seen during the War. Unfortunately he died after 4 days in spite of all resuscitation methods.

After about two months I was allowed to do some minor operations myself! It was during one of these episodes that I had an exciting experience. I was operating on a case of varicose veins. One of my fellow House Surgeons was also doing a similar operation in the same large operating room. I had nearly completed my operation when the other House Surgeon yelled at me to come over quickly to give him a hand to stem his patient's bleeding. I covered the wound of my patient and when I went across, I was met by a fine mess – blood everywhere and more blood welling up from the thigh wound. I took over, covered the wound with hot sterile packs and waited a few moments. Eventually I found the large varicose vein that my colleague had cut and failed to tie off. It was quite a tricky procedure as the main femoral vein was close by. But all went well and we decided to keep the incident to ourselves. However, the next day I was summoned to Mr Sangster's consulting room, and he had apparently heard about the operating drama. I was very relieved when he thanked and congratulated me, and then appointed me as his Surgical Registrar. Not only that, but he then suggested that I study surgery as my career. He advised me what books I should read. I took the train to Edinburgh one day and bought those books – second hand of course! I started studying surgery from there on and, after five months in Ayrshire, I decided to seek pastures new. I was having too good a time here when off duty and, instead of studying my surgery books, I went pigeon shooting with my twelve bore shotgun. I even took up riding again, accompanied by a friendly farmer's daughter, and managed some fishing in the River Ayr.

And so I applied for, and was appointed to, the post of House Surgeon at Larbert Orthopaedic Hospital, near Falkirk. This was again an E.M.S. hospital and it was here that I learnt traumatic and orthopaedic surgery in general. The Consultant Surgeon was a Mr Smillie – known to his juniors as the "wee man". After a year, I was promoted to being one of his registrars, with the responsibility of some operative surgery. He impressed me a great deal. The reduction of a broken bone had to be perfect. The re-alignment of a fracture with displacement of the bones had to be brought back to normality if at all possible. I also tried to attain that perfection and I learnt to treat all manner of fractures, for many of our casualties were men involved in mining accidents.

Mr Smillie, who later became Professor of Orthopaedic Surgery at Dundee University, was a pioneer in knee surgery, such as cartilage and ligament injuries. We, the House Surgeons, gave all the anaesthetics; there was no resident anaesthetist in those days. Mr Smillie was also a leader in hip replacement surgery and I assisted at many of these procedures. From time to time, we accepted convoys of injured armed forces personnel. Generally, they were soldiers (of all ranks), but a few were R.A.F. casualties. On one occasion, we were inundated with so many cases sent from North Africa that we worked through the night. I think I worked non-stop for 36 hours. I shall never forget two severely injured soldiers, one a sergeant and the other a private. Both had a leg in a Thomas splint, encased in Plaster of Paris from the foot to the pelvis. This was an excellent method of immobilising the injured leg and transporting the casualty back to the U.K. In each case, they suffered from a large wound in the thigh with an underlying fracture of the shaft of the femur. There was visible staining of the plaster, which was very smelly. The treatment was to remove the splint and the plaster, redress the wound and deal with the fracture. Neither man complained of pain, each seemed comfortable with the immobilisation that the Thomas splint afforded. It was when the Plaster of Paris was cut off and the splint removed that all of us were completely staggered. The size of the wounds in both men's thighs was huge. Furthermore, no attempt had been made to stitch the wounds, which were covered with gauze impregnated with Vaseline. The removal of the gauze revealed a wriggling mass of maggots! When these too were removed, it was quite astonishing and remarkable to find that the wounds were clean and devoid of sepsis. The men were treated in the usual way and eventually transferred to hospitals near their respective homes.

When I reflected upon this incident, I often wondered if any research had been done on the use of maggots in wound treatment. Can maggots have a built-in antiseptic and healing compound? One's imagination was apt to wander or wonder. Today, of course, such treatment is well known to be beneficial.

In the early days of the war, the mortality from septic wounds was considerable. The beginning of my career coincided with the introduction of antibiotics, the first being the sulphonamides. I recall with sorrow that one of my fellow medical students was afflicted with a severe boil on his cheek. The staphylococcal infection from this spread to the brain, with fatal results. A few years later this might have been successfully treated with the more advanced type of antibiotic.

When I was at the Larbert Orthopaedic Hospital in 1943, penicillin was being introduced. Its efficiency in wound treatment, and infections in general, was being investigated. I remember that, about twice a week, a small conical flask with a cotton wool stopper arrived by ambulance from Edinburgh. We used to study and photograph wounds treated with and without penicillin. It was truly amazing to see how wounds sprayed with some of this yellowish penicillin fluid healed so much more quickly. These patients were fortunate in not enduring as much pain and discomfort as those not treated with the new "wonder" fluid. I feel honoured to have been involved in this pioneering venture.

I continued to study a great deal, reading surgery in all its aspects. Quite often, I remember asking the night sister to wake me up at 4.30a.m. so that I could do an hour's reading before going back to bed again. In the residency my room was small - I did have one easy chair but my bed was my desk. My fellow residents were all good friends of mine. One evening after some alcoholic celebration, one of my great pals, Mohammed, an Egyptian chap, let off the fire extinguisher. As you can imagine, this resulted in a dreadful mess everywhere. Eventually, I managed to grab the exploding extinguisher and shoved the nozzle end down the lavatory. We all had to clean up afterwards and share in the cost of a replacement extinguisher.

Doctors and nursing staff both worked very hard, the doctors frequently ignored off-duty times, whereas the nurses usually did not, sticking to their on-duty rotas except in an emergency. One day, during a miners' strike, which was causing some concern generally, we had to deal with a recalcitrant miner who had fractured his spine. He was aggressive and awkward, appearing totally ungrateful for all our efforts to help him. Our treatment was to suspend the man by his feet from the ceiling, with his chest resting on the operating table. This forcibly extended the spine and, by leaving the patient in this position, together with added gentle hand pressure, we exaggerated the extension. This procedure helped to reduce the impacted vertebral body fracture, probably the 3rd or 4th lumbar vertebra. When he continued to be unpleasant, we announced it was our lunch hour and that we would come back later to finish off with a Plaster of Paris cast to his back. We all left, shut the operating theatre door and watched him through the door window. It was not long before he started to yell for help in no uncertain language. Of course, we quickly returned and completed his treatment. We wanted to demonstrate that medical staff were committed outside duty hours to finish a task and that the welfare of patients was a priority in all cases. He was a chastened man after that and apparently related this story to his fellow miners, praising the nurses and doctors for their skill, patience and forbearance!

Chapter 4: Hospital Posts

After I had spent two and a half years at Larbert Hospital, I wanted to move on to pastures new. I still hankered after general surgery, though I must admit I enjoyed my training in orthopaedics and, in particular, traumatic surgery. I decided to go to England, as more jobs were available there. I applied for, and was appointed, Senior House Surgeon, and later Registrar, at Salford Royal Hospital. Normally, this hospital employed about ten house doctors but now employed only six, so we were all very busy. The residency was adequate and my room was commodious with a large wardrobe, dressing table and a table I used as a desk. It was very noisy, as my window looked straight on to the main road to Manchester, down which sped trams making a phenomenal clatter from seven in the morning until late in the evening.

The hospital and surrounding area were extensively bombed about a year before I arrived and the young consultant surgeon, who was my Chief, had been awarded the George Cross for his bravery in rescuing some of the survivors from the damaged buildings. I was indeed fortunate to work under him, for he taught me a great deal of general surgery, sometimes assisting me and also leaving me many operations to do myself. I was in charge of the Casualty Department. A steady stream of casualties came in and all had to be dealt with, some being admitted to hospital and some referred to an appropriate department for treatment. I quite often had over 100 patients to deal with at one session! At the end of most mornings, I had a few minor operations to do, using a local anaesthetic or a general anaesthetic with nitrous oxide and oxygen. The latter was administered by me, then the Casualty Sister took over - an exceptionally competent nurse - whilst I dealt with the operation. This was usually the opening of an abscess, often of the breast, but also of the hand, the latter being often very severe. One had to be a quick operator. Most of these patients received a large dose of penicillin by injection - a very painful procedure in those days. It was quite remarkable how these patients' wounds healed, thanks to penicillin therapy.

One morning I shall not forget, for the senior consultant surgeon, who had come out of retirement because of the war, arrived at the Casualty Department as an emergency. He was in considerable pain due to a kidney stone causing acute renal colic. I attended to him, gave him an injection of pethidine to alleviate the pain and handed him over to the Consultant Surgeon. I was asked to set up an intravenous drip of glucose-saline. To do this, I had to cut down on a vein in the forearm under a local anaesthetic. In those days, this was the only method of firmly securing the needle in the vein. I had only assisted this "great man" the week before at a major abdominal operation. At this operation, he had got half way through when he felt he was too tired to continue and asked me to complete the operation - what an honour for a young aspiring surgeon! The patient soon recovered and was sent home.

Early one morning, I think it was about 6a.m., I was awakened by a very odd rumbling sound, and my bed seemed to be shaking. I nearly fell out of it! It was fairly light, being spring time. Suddenly, I saw my wardrobe doors gradually swing open. It really was quite frightening and I felt the hair on the back of my neck rising. Then I realised that this was an earthquake, which did sometimes occur in this part of England. It really only lasted for a few seconds, but it seemed so much longer. I got up and went out into the corridor of the residency and met some of my colleagues who had been equally scared. However, it was a minor tremor and no damage was done.

We occasionally had convoys of war casualties. Some of these casualties required surgery immediately and others were referred to hospitals nearer their own home. One patient I recognised as a colleague whom I knew as a medical student and who qualified with me. He had a healed wound on his face involving the parotid gland, but the wound was leaking fluid from this gland. He also had a healing wound of his groin. He told me that he had been hit by a sniper. The bullet had entered and come out of his groin and in doing so had injured the main artery of the thigh - the femoral artery. He told me he managed to stop the bleeding by pressing on the profusely bleeding artery with his fingers, until medical help arrived. I managed to arrange an ambulance car to take him all the way from Manchester to Edinburgh Royal Infirmary.

A group of German officers were included in a batch of casualties. Some of them seemed very frightened and some rather arrogant. One particular fellow took my thermometer out of his mouth, broke it in two and threw it onto the floor. He was the last to receive any treatment! They were, of course, transferred to a P.O.W. camp hospital. A Russian soldier was among the casualties and he was not badly injured. He was a most jovial chap and became popular with all the staff and other patients. He was also a great help in general ward duties until he was declared fit and had to leave.

One of my jobs was to assess the disability of men and women from all three forces - the Navy, Army and Air Force. I was the Civilian Member of the Pensions Board and my opposite number was a Colonel. The extent of the injury, disability or illness of the individual determined the grade of pension due. Also, we had to decide if he or she was fit to continue service, albeit with a reduced level of duties. Anyone suffering from a psychiatric condition or a stomach ulcer, a cancer, bronchial asthma or middle ear disease was graded "E". In other words, discharged as unfit for further service. We suggested the appropriate percentage pension to be awarded if we thought the condition was entirely due to serving in the forces. As far as injuries were concerned, we again had to decide on the appropriate degree of disability and hence pension due. We, of course, referred carefully to the disability percentage charts as prescribed by the Ministry of Defence.

My work at Salford Royal Hospital was very demanding but, I must add, also very enjoyable and highly instructive. I learnt a wealth of surgical knowledge from my superiors. Every evening, I studied my textbooks on surgery, often working on until midnight. Quite often, my evening's revision was interrupted by the arrival of an emergency surgical problem. This sometimes required an immediate operation perhaps for an acute appendicitis, an ectopic pregnancy, a perforated stomach ulcer, a strangulated hernia or an injury from an accident of some sort. When this was finally dealt with, it was back to my books to study.

My tenure of office was for a year, and so, by the early spring of 1945, I decided to sit the Fellowship exam in surgery at Edinburgh. I found it very difficult, for it embraced such a large field of surgery. I failed this, but I thought I had done quite well in some aspects and I was determined to have another go! I arranged to go to Edinburgh, to stay in digs somewhere, and to take the advertised ten week course in post-graduate surgery in the autumn of that year. So I went home to my parents feeling rather disconsolate. I worked on my father's farm for a week or two and then managed to get a job for two months as the Casualty Surgeon at the Coventry and Warwickshire Hospital. The hospital itself had been completely destroyed during the awful bombing of Coventry, the Casualty and Out-patients Department was the only part of the hospital remaining. The doctor I temporarily replaced was taking leave to sit his Fellowship exam in surgery. He briefed me as to the job in hand and told me I was entitled to charge patients if they arrived for treatment after 7p.m. This arrangement boosted my income quite a bit! He demonstrated one of his pet procedures for dealing with a severe injury to a nail bed resulting in a large collection of blood, i.e. a haematoma, lying underneath. This was usually the big toe nail or sometimes a finger nail. He used a Black and Decker drill which was attached to a wall bracket with the fine drill head pointing vertically downwards. The patient's traumatised nail was then, very carefully, led to the fast rotating tip. As soon as the centre of the nail was pierced, blood immediately poured out, thus relieving the tense haematoma underneath. I did use this method, but it frightened me, for if the drill accidentally went in too far, the underlying nail bed could be permanently damaged. I devised an easier method, more or less painless, and far less traumatic for the patient than facing the noise of an advancing high speed drill! My method consisted of easing a slender, sharp-pointed Bard-Parker blade into the side of the finger, just easing the blade under the nail between skin and nail. This was relatively simple as the tense blood clot had raised the nail off the nail bed. Perhaps I should have published this "minor op" procedure but never got around to doing so. The field of surgery is full of inventive ideas, some accepted and many discarded.

I was surprised at the scope of surgical cases that I had to treat. Many were emergency abdominal cases, which I referred to the Coventry Hospital, now relocated to a large convalescent home. I had quite a few fractures to deal with, wrists and ankles, etc many requiring manipulation to reduce the deformity back to normality. I often gave a local anaesthetic or intravenous pentothal. Occasionally, I used a general anaesthetic with ether and trilene. I administered this myself and, when the patient was "under", I handed the face mask to the sister-in-charge and then dealt with the minor operation. Perhaps this was to open a breast abscess, a septic hand or even an in-growing toenail.

My digs were not very far from my place of work and I got to know my way about Coventry quite well. I was appalled at the devastation of this fine city by the German bombing in the previous year. I walked through the ruins of Coventry Cathedral and wondered at that time if another could be built in its place. I was so very pleased to find quite nearby a small and very old church, which was relatively untouched and standing in its aura of defiance. I went into this church and immediately had a feeling of reverence. I sat in one of the pews and thanked God that this beautiful building had been preserved.

I was still working in Coventry when the end of the Second World War in Europe was declared on May 8th, 1945. I shall never forget the celebrations held that day and night, and for the following few days. I was on duty, of course, and had to deal with a large number of casualties - generally wounds, bruises and fractures. I was sorry to leave Coventry; the experience I gained there was invaluable.

Chapter 5: GP Locum & Fellowship

It was back home again to Scotland for a while, but shortly afterwards I was offered a locum in general practice. I accepted this with alacrity and, for the next four months, I was based in a fishing village - Lossiemouth - not far from my home on the Moray Firth. The nearest hospital was a small provincial one in Elgin and the nearest large one was in Aberdeen, sixty miles away. The senior partner of the practice, Doctor Tom Brander, was an elderly man, and I covered for his colleague who had elected to go abroad on a prolonged holiday. I lived in Tom's house, and he and his wife, Peggy, were very hospitable. I thoroughly relished this part of my medical career, so much, in fact, that I developed a feeling that if I was not successful in gaining my Fellowship at my next sitting, then general practice would be my choice of career.

I found, in this short time in general practice, that I enjoyed talking to and meeting patients from all walks of life, getting to know families and their varying health problems. It gave me great pleasure to prescribe, and to make better, many ill adults, children and babies. I did a lot of domiciliary midwifery, as the senior partner usually opted out. Chloroform, on its own an excellent anaesthetic, or, together with ether, the "C.&E." mixture, was often used. In one case, my Chief was delivering a baby with forceps - a difficult birth. I gave the anaesthetic – chloroform, of course - in a small cottage bedroom. I had to kneel at the top of the bed with the patient's head between my knees. The old doctor then asked me to take over, as he lacked the strength to exert a good traction on the forceps and he was gasping for breath. I'm glad to say that all went well.

One obstetric case I remember vividly. This was a call in the early hours of the morning to deliver a first baby. I cycled there, as in those days I couldn't afford a car. When the baby was eventually delivered, the new father was so pleased that he gave a very generous glass of whisky to both me and the midwife. Whisky was not my tipple, for I preferred beer, but I could not refuse and I had to drink it all down. There was no opportunity to pour some of it into the nearest flower pot! By the time I had finished, I was somewhat intoxicated and certainly very happy. I got on to my bicycle but fell off three times, so gave up and walked a good mile, mostly uphill, back to my partner's house.

Whisky was frequently used as a medicine by the elderly doctor. He supplied me with a whisky flask for those night calls dealing with a crying baby. If nothing particular was wrong, I gave the baby a tiny "tot" and I must say, this worked like a charm! I was instructed to charge seven shillings and sixpence (7/6) which was two shillings and sixpence (2/6) more than the usual call-out fee.

In the consulting room, which was part of the practice house, I was surprised at the number of minor operations that had to be performed, usually requiring a local anaesthetic, often a cold spray of ethyl chloride. Opening septic fingers or hands had to be effected over the waste-paper basket - a far cry from my hospital surgical training. But it was remarkable how these patients got better with the help, of course, of the judicious use of penicillin therapy. I made many friends in this seaside fishing town and was sorry in many ways to leave. Dr and Mrs Brander very kindly suggested that I could return to them if I ever thought of giving up my ambition in surgery.

Back in Edinburgh, I very much enjoyed the post-graduate surgical course. In order to make some pocket money to pay for my digs etc., I managed to tutor the occasional medical student. I was also very fortunate in obtaining an appointment at the Anatomy Department. Here, I was a Demonstrator and part-time lecturer. I remember two men who had been medical students with me. They had both failed their exams twice and were subsequently called up for the Army. Now they had returned to study medicine again. I had to give them both an oral exam on the part of the body they were studying. One failed and the other passed. The one who failed took the oral again the following week and this time I passed him. I expect they both eventually graduated as doctors.

I sat the Fellowship examination again in January 1946 and this time my knowledge of surgery was considerably greater than when I had failed in June of the previous year. It was indeed an exacting test, and as it happened, my great friend, Leonard Leitch (no relation), also finished his examination for the Membership of the Royal College of Physicians (M.R.C.P.) on the same day. That evening, we went together on a glorious, mad pub crawl. How we ended up in our respective digs, I just cannot remember! I think we danced the Palais Glide with some other candidates somewhere in Princes Street. I learnt later that Leonard was successful in gaining his M.R.C.P. and eventually became Consultant Pathologist at Dunfermline Hospital.

I went home again after the exam and I shall never forget a wonderful telephone call from my distant cousin, Mr Eric L. Farquharson, two weeks later, who said, "Let me be the first to congratulate you… you have passed the exam and are now a Fellow of the Royal College of Surgeons of Edinburgh". My father opened a bottle of champagne and we had such a happy family celebration. I was no longer a "Dr" but, as a surgeon, I was a "Mr". So my next task was to earn my living and find a job somewhere, and jobs were pretty scarce at that time.

One does hear amusing stories of candidates failing the Fellowship examination. One applicant, in the surgical pathology part of the exam, was given a narrow-shaped bottle containing a large tube split along its length. He turned the bottle this way and that, and when asked to describe what this specimen was, he suggested it was probably an elephant's ureter, i.e. the duct running from the kidney to the bladder. This was not the case at all! It was actually a specimen of a grossly distended human ureter - a hydro-ureter - due to a congenital abnormality and chronic obstruction between the bladder and the kidney.

Another candidate was presented with a female patient in bed on a surgical ward. He was asked to examine her, describe what he saw and to give the diagnosis. The candidate said the woman looked very pale and in a state of shock. He thought there was a cancer of the right breast present and that this had spread to the liver, making the latter organ enlarged. He also concluded there was fluid in the abdomen. "Well, what is the diagnosis and prognosis?" asked the examiner. The candidate announced that the patient was very ill, the outlook was poor and that she was suffering from an inoperable cancer of the breast. In exasperation, the examiner threw his books on the bed and shouted, "Can't you see, the blessed patient's dead???"

Chapter 6: A Young Surgeon

I applied for two posts, one at Ashford in Kent and the other at the Norwich Hospital. I was short-listed for both and eventually attended interviews at each hospital. I first attended the Norwich Hospital and was appointed. They did not offer me much of a salary and I told them so. It was a daunting experience, for I sat at a table in front of about twelve hospital Consultants. They decided to appoint another applicant who accepted their terms. I then attended the interview at Willesborough Hospital, Kent and I was interviewed by just four Consultants. They appointed me as Assistant Surgeon and with a good salary, which I accepted. Two weeks later, I had a letter from the Norwich Hospital saying that their previously appointed candidate had turned down their offer and they would be pleased to appoint me at the salary I had requested. I had pleasure in informing them that I had accepted another post!

Thus, my first post as a full-time qualified surgeon was at Willesborough Hospital in Ashford, Kent. It was at this small, provincial hospital that I really learnt, loved and revelled in the art of surgery. Thanks to my Consultant supervisors, all of whom lived in Folkestone, twenty miles away, I quickly learnt a wide spectrum of surgical techniques. My first love was to deal with traumatic and abdominal surgery. One Consultant was an expert in local anaesthesia and was kind enough to assist me at some major procedures. He had written a book on local anaesthesia and taught me his techniques and how to apply them, even when performing major operations such as partial gastrectomy and radical mastectomy. I did hernia repairs and other operations on elderly frail patients, using a local, spinal or caudal anaesthetic. This was the preferred method to avoid the risk of chest complications which often occurred with the use of ether in general anaesthesia.

The Consultant in Gynaecology and Obstetrics was a most jovial and much admired surgeon. He also taught me a great deal of gynaecology. I did not really take to perineal repair procedures but did like the abdominal aspect. In fact, after I had assisted at two Caesarean sections, I was given the task of doing any others which followed. I also worked with the Ear, Nose and Throat Surgeon performing many nose and throat operations, especially the removal of tonsils by dissection and using a local anaesthetic. But I did not enjoy nasal operations, and ear operations like mastoids, terrified me. In the speciality of Ophthalmology, the Eye Surgeon also taught me how to operate for squints and other eye disorders. But again, this was not my field of interest as a surgeon.

In the course of my work, I quickly learnt that, as a doctor, one had to be emotionally detached, even if the patient was a friend or a colleague. This perhaps particularly applies to the surgeon. The patient lying on the operating table and deeply asleep under a general anaesthetic, has signed a consent form giving one permission to operate. The patient must have complete trust and faith in the skill and ability of the surgeon to restore his health. It is indeed a considerable responsibility. Surgery necessitates sound knowledge of one's speciality. It is a demanding discipline but often a rewarding vocation, for a surgeon can help patients return to normal health in a very personal way. Lord Moynihan defined the attributes of a surgeon thus:-

"He is a physician who operates. He is someone with a lion's heart, an eagle's eye and a lady's hand."

One particular operation was both memorable and exciting. The patient was a man who, some five years previously, had had a steel plate inserted into his thigh in order to consolidate and reduce a severe fracture of his femur. The steel plate had become eroded and was causing the patient considerable pain, thus it had to be removed. The surgeon was one of the two General Surgical Consultants and a very charming man, but he did not enjoy good health. He had undergone a lobectomy to remove one half of one lung for a cancerous growth just a year previously, and this caused an occasional fit of coughing

The patient was anaesthetised and the operation commenced. When the offending steel plate was exposed, it appeared badly worn and rusty. This was the usual legacy of using steel. In later years, steel was superceded by a non-corrosive metal such as titanium. My Consultant colleague started to remove some of the screws. This was hard work and he was finding the task very tiring. One screw was particularly difficult and as my Chief was trying to turn this, it suddenly broke and the patient suffered a massive haemorrhage. Blood spurted out of the wound and even hit the large overhanging operating light! It was both dramatic and frightening. The main blood vessel - the femoral artery - had been punctured. My Chief then had a fit of coughing and asked me to take over the operation. I partially managed to stop the bleeding by putting my fingers over the bleeding point and a tourniquet was immediately applied. I was now able to remove the screws and complete the operation. The patient was given a blood transfusion and, thankfully, he made a good recovery. During operations, it is always prudent to expect the unexpected.

My opposite number on the medical side was the Medical Registrar and Obstetric House Surgeon - Geoffrey de Kyser. He was a charming and delightful colleague, a good physician but, if he had a difficult obstetric case in delivery, he often asked me to assist him. His wife was Fanny Waterman, who was a wonderful pianist. She had her own grand piano brought to the hospital, but the only possible place it could be housed away from the wards, was in the post-mortem room! So when I had to do a post-mortem, the porter would move it to a small adjoining office and then back again. Fanny played her grand piano every day, often all day, and always beautifully. She eventually went on to run the "Young Musician of the Year" competition in Leeds under her maiden name.

For routine operations, and some emergencies, a Consultant Anaesthetist attended from Folkestone, and sometimes there was a resident Anaesthetist available. Many of the emergency anaesthetics, however, were given by two local General Practitioners and they were excellent. If no anaesthetist was available, then I gave the anaesthetic myself or asked Geoffrey de Kyser to help. We were a good team.

One of my jobs at this hospital was to perform post-mortems (P.M.s), which were essentially routine. Usually a P.M. was done to establish, beyond doubt, the cause of death, sometimes in the interest of research and to find out if anything else might have been done to treat the patient's condition. Consent to perform this had first to be obtained from the relatives of the deceased. I enjoyed this investigating aspect of pathology. I also performed post-mortems for the Kent County Council, which were mainly for the Police. It was necessary to identify the cause of death in accidents or in suspected suicide cases.

If I was at all suspicious of a murder, I referred the case to a Home Office pathologist. Occasionally, I had to attend Court to give evidence as an Expert Witness. One of my most harrowing experiences was when I had to do a post-mortem examination on six victims of an air crash which had occurred nearby. All died of multiple injuries but a careful list of all injuries had to be given in the pathological written report.

For my recreational activities, I played rugby for Ashford. We had some great matches, especially against the "Skinners" team - former pupils of Tonbridge School. Our team consisted of two or three GPs, dentists, solicitors, businessmen and police officers. Sometimes, I went shooting pheasant or partridge with GP colleagues, using my own 12 bore shotgun, kept in a wonderful old leather gun case. Besides housing my gun, it had a grey metal tray with lots of compartments. This probably used to contain equipment for making gun cartridges. The case was originally owned by my great grandfather.

One day, Geoff and I decided to celebrate his wife's birthday. So, four days beforehand, I went into a nearby wood where I had heard a cock pheasant calling and where I thought I would be safe from prying eyes. I stalked this bird and, when it flew up, I shot it with my 12 bore. Annoyingly, it fell into some thick bramble bushes. Just as I started to look for the bird, a rabbit ran out and I shot that too and picked it up. Suddenly I was startled and terrified to hear a gruff voice shouting "Who's that shooting?" I well knew I was poaching but tried to appear gormless and baffled at his belligerence. "I've just shot a rabbit with my second shot - I missed it with my first." I held up the rabbit. "This is private land and who are you?" he demanded, "Oh dear, I do apologise. I am a doctor from the hospital and my friend and I wanted a rabbit for his wife's birthday," I replied. He accepted this with a bad grace and asked to see my gun. This request I refused, "This is an old gun belonging to my father who is a farmer like you." He grunted and left and I returned hastily to the hospital with my precious gun and the rabbit. Geoff thought my adventure hilarious and was delighted to have an unexpected rabbit supper. Without telling anyone, later that evening, when it was quite dark and nightingales were singing, I crawled back into the wood and with my doctor's pen torch managed to find that pheasant. How fortunate that it had fallen into the brambles after all! I made a hasty and silent retreat with a fast beating heart. Oh Boy! In five days' time, that cock pheasant roasted would be scrumptious.

I worked at Willesborough Hospital for nearly three years. A month's annual holiday was allowed and this I spent mostly with my parents in Scotland. I still remember travelling overnight from Euston to Inverness, in an empty compartment if I was lucky, and often being pulled by the "Flying Scotsman" - what a beautiful steam engine that was!

As I was now a fully qualified surgeon, I thought I might help my father. He owned a herd of Large White Pigs which won many prizes at agricultural shows in Northern Scotland. He often judged at those shows and even did so at the Royal Highland Show in Edinburgh. He occasionally had a pig which suffered from a rupture of the groin - or inguinal hernia. Such a pig would not thrive to reach the usual weight of 20 score pounds, which was required by the bacon factory. To save them from the knackery, with my father's approval, I operated on these ruptured pigs under a local anaesthetic, using a set of surgical instruments borrowed from my hospital - I blessed those kind Theatre Sisters! The pig was held up by its hind legs and gripped tightly by our pig-man, Alfie. He had been a P.O.W. at a camp near Forres and had married a local girl after the war. I am pleased to recount that each pig, although in shock for a while after the operation, was soon drinking copious amounts of water and eating food. Post-operatively, each pig was injected with a large dose of penicillin. In every case, the pig attained the magical weight of 20 score. I also operated on a cow, removing a huge fatty tumour, a lipoma, with success - especially as she started producing milk once again. It was quite an experience!

I acquired my first car, and became the proud owner of a Morris two-seater Tourer, registration number CSP 166. The windows were plastic and the hood was apt to leak! The winter of 1946-47 was a very severe one. Just at this time I was courting a General Practitioner's daughter, whom I eventually married in July 1947. She lived in Folkestone, and motoring the 20 miles there at night was quite a hazard. My windscreen would freeze over and the only way to clear it was to smear glycerine on the outside. Not many cars had heating in those days. I well remember the late afternoon that I proposed, and was accepted, when walking through a wood. I had my trusty 12 bore shotgun with me and was supposed to be shooting rabbits and wood-pigeons. We went to a local pub she knew well and the proprietor presented us with some champagne. Whilst we were celebrating, a telephone call came through from the hospital from Dr de Kyser - would I please come quickly to help with a difficult obstetric case. I immediately left for the hospital. I had to deliver the baby by a lower segment Caesarean section. All went well and I returned to the pub with my now fiancée, Betty, to continue our celebration and finish the champagne. This whole episode was indeed a romantic drama worthy of Mills and Boon! When we were married in July, we shared the residency together with Dr and Mrs de Kyser and an anaesthetist.

A few years later, whilst my wife and I were on holiday at a hotel near Loch Ness, Taggart, our Staffordshire Bull Terrier, was run over. Taggart had sustained a badly fractured femur and we took him to the local vet, who rather callously said that the only treatment was "to put the dog down". I did not accept this. We set off for the Inverness Royal Infirmary. The staff were highly amused, but extremely helpful - they had never had a dog as a patient. Can you imagine this happening today?! Taggart was x-rayed to confirm the position of the fracture. I then sedated him with Pethidine, manipulated the bone into a reasonable position and applied plaster of Paris - a hip spica reinforced with cramer wire. This wire splint was left to protrude below the plaster. I covered it with a piece of rubber so that Taggart could hobble about. After four weeks, we took him to another vet, who was amazed to see such a plastering contraption. He refused to remove it so I borrowed his saw and did it myself. Fortunately, all was well and the fracture had healed, leaving Taggart with just a slight limp.

My tenure of office was shortly due to come to an end and I had to think of another move. I was offered the position of Assistant Surgeon at Folkestone with a view to future Consultant status. Unfortunately for me, the National Health Service had just been introduced and all new consultant posts were shelved and indefinitely postponed. This was a blow to my aspirations.

However, I applied for and was appointed to the post of Orthopaedic Registrar at the Chester Royal Infirmary. Because of my interest and training in abdominal surgery, the two General Surgeons at the hospital asked me to be on call for emergencies on three or four nights per week. They both sometimes asked me to assist them whilst operating on a private patient. I was paid well - any contributions were gratefully received! Likewise, the Consultant Orthopaedic Surgeon paid me if he had a private patient to deal with. He also gave me some of his insurance work. This involved patients who had suffered injuries whilst at work such as fractured bones, serious or otherwise, and who were making a claim for compensation. On the strength of this, I bought a portable typewriter and a booklet on the art of typing. I learnt to use all of my fingers, but always had to watch the keyboard. That typewriter paid for itself many times over. When I had to write up my Income Tax Return, I was worried about how much of this extra income from insurance reports I should declare. I made an appointment to see an Income Tax Inspector. I told him of my dilemma. He really was very helpful, for he simply said that what I earned was well deserved and not to mention any figures. Apparently, he had learned that I had been very kind to his wife when she had brought her screaming child into the surgery. My Chief could not abide shrieking babies and referred them all to me. That particular consultation had a lucky outcome!

My tenure of office at Chester Royal was for a year only. I was replacing the current registrar whilst he was on leave to take his Fellowship examination. My wife and I were lucky to live in his rented and partially furnished house. At this time, of course, rationing was still the norm - the aftermath of the war. Everything was in short supply. I was most fortunate in working at the hospital, for patients often generously presented me with eggs or milk and even the odd petrol coupon. During the Chester horse racing season, I treated a few injured jockeys. They gave me a few tips on which horses were likely to win! I thought I was "on to a good thing". But, no! I was about £50 up at one time but by the end of the season I was no more than £5 in credit!

During our year's residence in Chester, my wife attended many furniture sales. We used to go to the sale-room early, often on my way to work, and set a price on the pieces of furniture we fancied. Soon, our rented house was very well furnished, as my wife really was proficient at bidding, but we still had no property of our own.

Chapter 7: The Midlands

When I applied for another post - in 1948 - jobs of any seniority were very scarce. However, I did apply for a Surgical Registrar's post at a small provincial hospital in the Midlands. This was The Horton General in Banbury, Oxfordshire. Banbury was a bustling market town where Midland Marts had the largest pig and cattle market in Europe. Here, also, was the huge aluminium factory, Alcan. Banbury, of course, is well known for its association with the nursery rhyme "Ride a Cock Horse to Banbury Cross".

My contract with the Chester hospital finally ended and moving was quite an exercise, for our extra furniture had to be taken down to Folkestone where my in-laws lived. Motoring to this Midlands town, our Standard 8 Tourer car was packed full to the brim. The hood could not be put up as the ironing board stuck out from the back seat. Keeping this in place was an assortment of suitcases, boxes, odd clothes and, of course, our beloved Staffordshire dog, Taggart, sitting up in the middle of the packing. When we arrived at the town, I wanted to change into a suit and so I stopped at what I thought looked like a good hotel. This turned out to be the Police Station! However, a Constable directed us to the main big hotel, but here, no dogs were allowed in and so we continued our search. Eventually we found a small, very friendly establishment where I changed into my suit. I attended the interview at the hospital, was offered the job and accepted it. The salary was reasonable. We then motored all the way to Folkestone - what a long day, especially for Betty who was already three months pregnant. Unfortunately, I had to become a resident in the hospital and so my wife had to move back in with her parents in Folkestone. However, in her eighth month of pregnancy, her mother was unable to help in any way. I was given a few days leave and I collected Betty so we could motor all the way to my parents' farm, near Elgin. I then returned to my work at the hospital. One morning, I had just finished an operation when a telephone call came from my mother to say that Betty had just been delivered of a son and all was well. Over the moon, I had to contain my delight and concentrate on returning to theatre and continue operating to finish my list. Immediately afterwards, I celebrated with my friends, drinking strong coffee laced with a suitable amount of whisky. No time off for new fathers in those days! Our new son was to be christened Nigel. It was, of course, a difficult time for Betty in being so far away from me. But she was so courageous in accepting the situation.

Fortunately, I was soon appointed Senior Surgical Registrar and with a much enhanced salary. There was no security of tenure in those days, even at this level of one's career, and I knew that my term of office was only for four years. In time, I would again have to look for another post. I worked very hard and was on call six nights per week for all emergencies - general surgical and orthopaedic alike - but I thoroughly enjoyed it all. Also, I had found a flat to accommodate my family, so our happiness was complete.

I lectured to the nurses on anatomy, physiology and surgery and I was also appointed as an examiner to the Royal College of Nursing. The latter post I continued for seven years. At this juncture, please allow me to digress. In following my love of general surgery, it is surprising to consider the breadth of surgical problems and the variety of options confronting me. I was, at all times, encouraged and assisted by my Chief, the Consultant Surgeon-in-Charge. I dealt with all traumatic surgical problems, wounds, simple and compound fractures and even the occasional fractured neck of the femur. In those days, this involved simply driving a triangular-shaped nail - the Smith-Peterson nail - across the fractured femoral neck.

Head injuries were often difficult to treat. In general, masterly inactivity was recommended. I do remember three cases. One was a man injured whilst playing cricket. He had been hit on the temple by a cricket ball. The x-ray revealed a fracture of the left side of the skull. He complained of severe headache, became badly disorientated and partially unconscious. This was only temporary, but we diagnosed an increasing blood clot due to haemorrhaging from a damaged artery at the fracture site - the middle meningeal artery. He was referred to a specialist head injury unit at the nearest teaching hospital some thirty miles away, The Oxford Radcliffe Infirmary. I later heard that he had recovered post-operatively but that the outlook was uncertain.

The other two patients were both unconscious upon arrival at hospital. Each x-ray showed, again, a fracture in the temporal area. Haemorrhaging from the middle meningeal artery was diagnosed. Immediate surgery seemed the only treatment to save life. I opened the skull at the fracture site. In one case, the haematoma, or blood clot, was huge and the bleeding point could not be found. The wound was packed with warm gauze, the patient, remarkably, recovered, only to die three or four days later. Similarly, in the other case, I found a developing blood clot at the fracture site. This time, however, the clot was sucked out and the injured blood vessel secured. Remarkably, and dramatically, the patient suddenly partially recovered on the operating table. He had been kept under very light general anaesthetic. This patient was eventually discharged from hospital. He was lucky!

Any emergency abdominal problem was also my responsibility for both diagnosis and treatment. Just occasionally, when a difficult and critical night emergency arose, I would telephone my Chief. He gave advice but often came in and operated on the patient himself, assisted by me. We still had to work the next day - there was never any thought of asking for overtime pay or for being paid extra for night work or for working during unsocial hours. This applied to all hospital doctors, be they House Physicians, House Surgeons, Registrars or Consultants. The only way to gain experience and increase expertise is to work hard and for long hours. Hard work, after all, is the very elixir of life! I frequently compared "notes" with my junior colleagues and we all mostly enjoyed our work - medicine was our vocation, our chosen way of life, and we were happy. Some of us hoped to gain a good testimonial from our "Chiefs". This would help lead to another good post and one day, maybe, to a consultancy in the speciality of one's choosing. Such an ambition, of course, involves great dedication and sacrifice.

It would be remiss of me not to give due praise to my nursing colleagues. Their profession was likewise a vocation. A good Theatre Sister was invaluable to a surgeon. The treatment and recovery of patients was always dependant upon the skill, patience and encouragement of their nurses. Frequently, they too worked extra hours, without recompense, in order to complete a task or during an emergency. As a young doctor, I quickly realised that I could learn a great deal from the nursing profession.

I must also pay tribute to my wife, Betty, who was always a tremendous support to me throughout my long career. She endured interrupted nights and she rarely complained when social engagements had to be cut short or even cancelled because of off-duty calls. When a patient telephoned she was always helpful and reassuring. Sometimes a caller would be over-wrought, but occasionally could be demanding and aggressive. Betty handled each with tact and sensitivity – the perfect doctor's wife. This, of course, applied equally to my partners' wives and I am sure to all doctors' wives or husbands. "They also serve who only stand and wait."

On Christmas Day, it was the custom in many hospitals for the doctors to dress up and carve the turkey on the wards. This was enormous fun for patients, nurses and doctors alike. I always wore a tartan Glengarry, a tartan scarf and waistcoat. The nurses made me a skirt from a white theatre gown. My Chief was likewise dressed up, but for a hat he wore a small bedpan! We carved the turkey, usually a very large one, in our male and female surgical wards. The medical wards were similarly served by house doctors or by the GPs who were on the staff of the hospital, the so-called S.H.M.O.s (Senior Hospital Medical Officers). Afterwards, the staff would retire to the Sister's office and eat mince pies and Christmas cake, washed down with tea, coffee and liberal amounts of sherry and wine. It was extraordinary, but very rarely did an emergency spoil our party.

I took part in many out-of-hospital activities. I joined the local St. John Ambulance Brigade and eventually, many years later, became their President and was awarded the Order of St John (O. St. J.). I have a letter from the Secretary-General for the Order in which he states "Her Majesty the Queen, the Sovereign Head of the Most Venerable Order of St. John of Jerusalem, has been graciously pleased to sanction your promotion to the Order of St. John". This honour was conferred on me by the Commissioner of the Order at the headquarters of St. John at St. John's Gate, Clerkenwell, London. I joined the local Caledonian Society which had just been set up. My wife and I took part in many sessions of Scottish country dancing. I was elected the Society's third President which I accepted with great pleasure. I had very many good friends in this town. Also, I made many wonderful friends of my GP colleagues, many of whose patients I had operated on.

Chapter 8: GP Surgeon

When the end of my four year term of office approached, I wondered what my next move would be. There were very few General Surgical Consultant posts advertised and those few usually at large teaching hospitals. However, I was short-listed for three and went for interviews, only to find that the current Senior Surgical Registrar was more or less already appointed. I mentioned my problem to one of my closest general practice friends. I said that I had been so very happy in my short experience as a GP and wondered whether this, after all, was the way forward - my destiny. Shortly afterwards, I was invited to join his Group Practice of five doctors in Banbury. I was to be the sixth as an assistant, with a view to partnership. I conferred with my parents and with my father-in-law, who was a General Practitioner. I accepted this post and it was indeed the correct decision to make. In the meantime I had been offered a job as a Consultant Surgeon by the Colonial Office in Tanzania. This I turned down.

Two weeks before I finally left my hospital job, my Senior Consultant Surgeon went on holiday. During this time I had a "field day". I performed a great deal of surgery including many major surgical procedures. That was my swan song as a full-time surgeon!

Within two weeks of joining the Group Practice, I was appointed part-time surgeon to the local hospital where I had previously, of course, been working full-time. My job there was to do all the emergency surgery on one night a week and to have my own operation sessions on one day a week. This latter session seldom involved any major operations. I was really a very fortunate doctor as I was now a GP Surgeon! On Monday nights, I was on call at the hospital for surgical and traumatic emergencies. It quite often happened that I would be called to see a patient at his home with an abdominal emergency and later operate on him that very same night.

One of my first jobs when I joined the Practice was to attend a private patient of the Senior Partner. She was a charming titled lady who lived 23 miles away. She had fallen off her horse and cut her head badly and, as a surgeon, I could deal with this accident. She had a deep laceration on her scalp and had refused to go to hospital. However, I enjoyed this type of surgery and stitched up the wound using a local anaesthetic. I attended her twice more to re-dress the wound and then, finally, to remove the stitches. I wondered what my Senior charged for this treatment. Nothing was paid to me as I was not yet a partner in the Practice.

I remained an assistant for about six months and then bought a share in the Practice. After a further three years or so, I became an equal full partner. Work was jolly hard with very long hours. I had only one full day and night off per week and, for the rest of the week, was on call for 24 hours a day. Night work was demanding and often very tiring for one still had to work the next day. But I became used to it and night calls often gave that extra "buzz" of excitement when one was confronted with an unknown problem. As the years passed, GPs were eventually paid an extra fee for night calls, the proverbial form having first to be completed and signed by both the visiting doctor and the patient. Any income, including private patients' fees and hospital work was all paid into the partnership and equally shared. Hospital work helped a great deal towards one's pension, index-linked, and likewise shared equally.

As a Junior Partner, I was given some odd visits to do, which, I suppose, reflected my position in the pecking order. One call was to a village five miles from the surgery, the patient being a child with suspected measles. I had to walk across two fields where cattle were grazing, and in one field there was a very large bull. This, luckily, was a Hereford and, being a farmer's son, I knew it should be pretty docile. But I nevertheless kept close to the fences, just in case! Then I had to cross the main railway line to finally reach the house, which was next to the canal lock. I confirmed the diagnosis - the boy was quite comfortable - and started to make my way back to the surgery. Unfortunately, I was caught in a thunderstorm and my clothes got so drenched I had to return home to bathe and change before I could resume my house calls.

I had another canal lock-gate house call, this time an emergency. The house was only half a mile away from the road and I was told there would be a bicycle waiting for me under the nearby bridge. I can tell you, it was quite perilous cycling along the narrow tow-path with medical bag in one hand, trying to keep from wavering and toppling over. I eventually reached the house and found that my patient, a boy, had appendicitis. I called the ambulance and this, having a four wheel drive, managed to cross two hilly fields.

Returning home, I was in no hurry and I chose to walk with the bicycle this time!

The other two calls were by horse! The first was to a remote farm house. It was arranged for me to go by horse and cart. When I arrived at the gate, there was the transport but, alas, no driver. It turned out that he was in the next field hoeing turnip seedlings. So, up I jumped and took the reins. The horse, Fred, was slow and cantankerous and, when I swore at him in my impatience, he stopped altogether. He had been brought up politely! I called here three more times and grew quite fond of old Fred. Years later, the farmer had a proper road built.

The second visit by horse was again over fields to see an elderly farmer - a delightful and well-spoken character. This time a 16-hand riding horse awaited me. Boy, was I glad I had been brought up on a farm! The horse was saddled but with no stirrups. I am glad no one could see my attempts to mount her from the gate, with medical bag in one hand, of course. The patient had pneumonia and he refused to go to hospital. I attended him several times and made sure that stirrups were provided so I could canter slowly to the farmhouse. Each time I called, the farmer's wife gave me a cup of tea, a slice of cake and a tot of whisky. On my final visit I was presented with a whole bottle of whisky.

I enjoyed motoring and on average covered 1,200-1,300 miles per month. I became a member of the Institute of Advanced Motorists (I.A.M.) in 1955. I took the gruelling test for this in Birmingham. It lasted for more than two hours and I was delighted to pass the first time. I was supposed to get a reduction in my motor insurance premium but, just because I was a GP, my request was turned down. I put this rule down to the fact that a doctor can be called out in any weather and at any time, day or night. At least I had the pleasure of fixing my official I.A.M. badge to my car radiator!

I had many exciting car journeys. I expect I must have been in my car at one time, every hour and day of the year, even on Christmas Eve, Christmas Day and New Year's Eve. One had to deal with many hazards. In the winter months, in case I got stuck in the snow, I always carried a spade, some wire netting and a sand-bag in the boot of my car. I also always had a bar or two of chocolate, a few biscuits and a bottle of water and a can or two of beer. And of course, I could not be without my first-aid case, intravenous glucose saline equipment and a good steel crowbar. This latter tool helped release the odd accident victim out of a car. Motoring on icy roads and in snow was very tricky. I found that it helped to keep one's steering wheel moving a little from side to side on snow-covered roads.

I dreaded driving in thick fog - a dangerous pastime at the best of times. It paid dividends to know the road ahead and all of its corners. In thick fog, I sometimes had to drive at no more than 10 m.p.h. with my left hand on the steering wheel and the right holding my door slightly open so that I could lean out and follow the "cats' eyes" or white lines in the centre of the road.

I was fortunate not to be involved in a serious accident and my licence remained unblemished throughout my career. I did, however, have one very frightening experience. Whilst motoring in a housing estate, I stopped my car behind a lorry which indicated that it was turning right. This was also where I was going. Suddenly, the lorry started to reverse into me. I sounded my horn and shouted and even gave piercing whistles with my fingers - a trick I had learnt when a boy. The lorry kept coming and the back of it came over the bonnet of my car and eventually came up against the side struts of my windscreen, shattering it. Fortunately, some workmen ran up to the lorry driver to tell him to stop. By this time, my car had been pushed back about 20 yards and the back of the lorry was very close to my head. The workmen helped me out of my car and gave me a cup of tea from one of their flasks. I was pretty shaken. I got a taxi driver to take me to my surgery where I arrived about fifteen minutes later. The lorry driver was duly charged and convicted of careless driving. The day's work had to go on but I had a good stiff whisky when I got home - the water of life - "Uisge Beatha", not omitting the Gaelic toast "Slahnja vah, Slahnja vor", meaning "Great health, Good health!".

Chapter 9: Work in the Community

For the first fourteen years as a GP, my family lived in houses owned by the Practice. Each partner was responsible for the surgery attached to his home. Our first house had its surgery in the basement, six stone steps off the outside pavement and the main road through the town. The large waiting room was heated by an anthracite burning stove which also heated the water. One early morning, three large bags of bone meal were left outside the surgery door. These had been sent down from Scotland by my brother who, besides managing the family farm after father's retirement, was a leading businessman in the area. He owned the Knackery in Elgin. It processed any dead animal found in a field as well as offal from slaughter houses. On one occasion, the factory dealt with a dead whale which had been washed ashore in Aberdeen. Two of the by-products were tallow - used in soap making - and bone meal. Well, I carried the three heavy sacks into the waiting room to keep them dry and then went into my consulting room to commence the surgery. I awaited my patients as usual and could not understand why the attendance was so low that morning - I only had 2 to 3 patients instead of the usual 20 to 25. It was only when I went back into the waiting room to close the outside door at the end of surgery that I found the heat of the stove had caused the bone meal to stink to high heaven! However, thanks to my brother's generosity, my garden bloomed as never before, in both flowers and vegetables!

From one of my patients, I acquired an upright piano which stood in the waiting room. By good fortune, another patient, who lived with his brother and sister, had been a concert pianist. He often used to come and play on the piano, most beautifully, to the delight of those waiting to see me - there was always a full surgery then!

This particular surgery house had a long, narrow garden at the bottom of which I always kept a few hens in a small hen-house. The eight or nine hens had a deep litter of leaves, which I collected in the autumn from nearby woods. The hen-house was covered over with corrugated iron sheeting. The hens, always brown leg-horn pullets, were kept for a year, then a fresh lot bought. The deep litter, now a year old and very rich was then spread all over the garden. One year in early autumn, I was staggered to find a large crop of lovely mushrooms growing! I was the only GP out of 12 in the town to keep hens and my partners and staff, although amused at my eccentricity, had the benefit of eating new-laid eggs!

Later, we moved to a larger practice-owned house adjoining our main surgery, which had the benefit of six consulting rooms. Here, the partners could work together in close co-operation. My daughter was born at this time - on May 1st 1955 - when, unbelievably, we had the heaviest snow-fall of the year. This was especially joyous for us as Betty had previously suffered two miscarriages. Now we felt our family was complete. In typical male fashion, I celebrated by buying a new Cortina, one with a pink roof! Nigel, in typical brotherly fashion, was much more interested in the car than in his sister.

On May 5th, one of my patients was delivered of her fifth child at 5.55a.m. in the local maternity home. I wrote this up in a letter to the local weekly newspaper as "All The Fives" - the fifth child on the fifth day of the fifth month of the year '55 at 5.55a.m.!

I tried to limit my surgery consultation times to around five minutes per patient. Sometimes this was less, quite often more, depending upon the patient's condition. Mentally ill patients often required careful and prolonged treatment and I frequently referred such cases to either a psychiatrist or to a psychologist. Of course, I had my fair share of hypochondriacs and these patients needed tactful handling. I once read a report of an epidemic of hypochondria which happened in Singapore. Apparently, hundreds of Chinese men began to believe their penises were retracting into their bodies. Rickshaws could be seen hurrying to doctors' surgeries, each containing a terrified passenger with friend holding firmly on to the unfortunate organ to "stop matters getting worse"! Fortunately, we have had no such outbreak here!

One of my first home visits was to a pub on a council housing estate. I knocked at the front door, walked into the porch and said "Good morning". I was confronted with a loud guttural voice, "Good morning, you bugger!" Rather taken aback, I huffily replied that I was not accustomed to being spoken to like that, and walked into the entrance hall of the pub - to be confronted by a most colourful parrot, who repeated his greeting to me once again! In this particular council housing estate, the houses were very old, and with no hot water tap. In some cases, the lavatory was outside in a back garden and shared between three or four houses. I used to catch the odd flea, which did not attack me as much as my wife! In one house, I removed my hat as I entered, but could not find anywhere to put it (I always wore a hat in those days, and always a suit and tie). I simply replaced my hat on my head and went upstairs to see the patient. Mighty few homes had a bathroom. In one house, I was surprised to see a crop of potatoes growing in the bath and doing well. All those houses have since been pulled down. Families were relocated elsewhere to much better accommodation with the result that the standard of living improved enormously. All new houses had hot water and proper inside bathrooms.

One family living in a council house, I knew well, for I delivered all five of their children. The children frequently ran about outside in bare feet and, really, they were remarkably fit, except for all the usual infectious diseases - mumps, measles, chicken pox and whooping cough. One visit I made at this house was during the winter. The fireplace was piled high with ashes. I had just finished examining one sickly child and started to write a prescription when there was a roaring noise from the fireplace. The chimney was on fire and soot came pouring down and into the sitting room and kitchen. I fled from the house with the family and then finished writing up my prescription from the safety of my car. There were, of course, volumes of smoke everywhere. The mother was not unduly worried, for apparently the chimney went on fire every two years, which saved paying for a chimney sweep! When I called the next day to see the young patient, all seemed under control, but the smell of burning soot would linger on for weeks.

From the same council estate, a choking child was taken by his mother to my own house, whereupon my wife urgently telephoned the surgery for assistance. The Senior Partner arrived and was impressed to find my wife by the front door, holding the child upside-down. Her prompt action had averted a tragedy as the boy had swallowed or inhaled something which had blocked his windpipe. When left upright, he choked and turned blue in the face. He was duly rushed to the local hospital, where the offending obstruction was quickly removed and the child made a complete recovery.

Dogs often presented a hazard to the visiting doctor. One evening, when it was pretty dark, I called at a house and was confronted by a large collie which growled and barked ferociously and then was obviously getting ready to leap up at me. Fortunately, I was carrying my large, heavy torch, which I used to hit the dog a resounding blow on its head, just as it jumped. It collapsed on the path and was well knocked out. I carried on and into the house. After dealing with the patient, I wondered about the dog but when I came out it was nowhere to be seen. That collie did not dare to attack me again! It was in the same housing estate that I was bitten slightly by a spaniel and on another occasion by a dachshund. Throughout my career, these were the only two breeds that bit me. When I encountered patients' dogs, I always held my doctor's case between myself and the animal and this worked well. Perhaps the smell of ether from my case was sufficient deterrent. In those days my syringes and needles, used to give injections or to take a specimen of blood, were kept in a special metal chromium syringe case and filled with ether. The ether leaked occasionally. In all the years of my practice, I never caused a septic injection. In years to come, all doctors were given, free of charge, sterile needles and syringes which were discarded after a single use. I never left my car unlocked whilst visiting. There was one exception to this rule. No one in their right mind would dare to enter my car when my Staffordshire Bull Terrier was in situ - sitting regally in the passenger seat and looking menacingly at any passer by.

One emergency call that is etched in my memory was the night I encountered a ghost. I was telephoned at home, well after midnight and asked to visit a seriously ill widower who was lodging at a vicarage, in a village some six miles away. I had never been there before and the vicar's wife told me to look for a door in a high surrounding wall and then walk to the house. Well, I found the door easily enough, but to my surprise the path led through the cemetery, then down twelve stone steps before finally reaching the manse. Fortunately, the moon was shining and I could just about find my way. The patient proved to have a severely infected gall bladder and I immediately made arrangements for hospitalisation by ambulance. After bidding goodnight to the vicar and his wife, I proceeded up the stone steps. Just as I reached the top, a scudding cloud obliterated the moonlight and I was left in total darkness. In trying to extricate my torch from my pocket, I tripped over, losing my grip on the handle. I must have looked a comic sight as I groped around on the path - my bag had fallen as well. However, I found the torch fairly quickly. As I switched it on, the strong beam shone straight across the cemetery, lighting up the church steeple - only to reveal a huge black spectre towering above the gravestones. This was surely a ghost and, with fast beating heart, I just gaped at this awesome "thing" before me. I also remember thinking to myself that there *must* be a scientific explanation. As I moved gingerly towards it, so it loomed larger, and slowly the truth dawned that I was seeing my own shadow! Feeling foolish, but relieved, I retrieved my bag and negotiated my way back to the outer door. But my fear had been real, and I did not wish to pay another midnight call to that particular vicarage.

I was particularly happy as a General Practitioner. Obviously, I missed my position as a full-time surgeon, but, as I still continued doing some surgical operations at the local hospital, I really had the best of both worlds. I was so delighted when my wife and I went to Edinburgh in 1992 to celebrate my 50th year of qualification and there met three of my fellow medical students! They also had their Edinburgh surgery Fellowship, had been GPs and did surgery part-time at their respective local hospitals. But surgery and medicine have advanced so much as the years have gone by that the era of the GP surgeon is fast becoming past history.

My previous Senior Partner and I were keen on attending GP Refresher Courses. In fact, all six of us partners took it in turns every two to three years to attend a fortnight's course. This practice continued when I became Senior Partner. I can clearly recall one course where there was a lecture on the art of hypnotism, given brilliantly by a doctor who was a member of the British Society of Hypnotists. I thought of exploring this further, but decided I already had plenty to do and to hypnotise a patient properly was really quite time consuming. I regularly went back to my Alma Mater and made copious notes after the lectures on my portable dictating machine. The notes were transposed later and a copy given to each partner. These courses brought us up to date with advances in all fields of medicine, essential for a successful practice.

Visiting old people was always worthwhile and I found Sunday a good day to call on them. Geriatric medicine was a large part of one's work and one might say that the doctor's job was not so much to keep patients alive at all costs but to keep them comfortable. By regular visits, one was sometimes able to spot and treat early signs of disease, be it cardiac, digestive, neuralgic or something more sinister. There is the story of the old lady who asked a friend who had just been in to see the doctor: "What was he like?" The reply came: "'E's not like a doctor, 'e's a gentleman!" The doctor's dictum written above the Mayo Foundation House is: "To cure sometimes, to relieve often, and to comfort always". This would apply especially within the field of geriatrics. Whilst I am on the philosophical aspect of medicine, may I quote: "A physician is a doctor who knows everything and does nothing. A surgeon is a doctor who knows nothing and does everything. A pathologist is a doctor who knows everything but always forty-eight hours too late"! On just the rare occasion, I was present when an elderly patient of mine died. I always found this a harrowing experience - the more so if I had been their family doctor for many years. It was particularly upsetting if a patient died of a heart attack, after refusing to be admitted to hospital as an emergency. I felt sad and helpless when all methods of resuscitation failed.

Fortunately, I was blessed with good health apart from my asthma, which rarely bothered me. I was also very allergic to cats. I kept myself physically fit by routine morning exercises, which I still do to this day. I often rose at 6.30 in the morning to do gardening before breakfast. I rarely missed a day's work. However, in mid career, I developed thyrotoxicosis, due to the overactive thyroid gland in my neck. This serious condition affected my heart and I was admitted to the Oxford Radcliffe Infirmary. For the first time, I learnt how it was to be on the receiving end of the medical profession and it was a salutary lesson. On my first night in the admission ward, an old man in the bed next to me died, with the resultant whisperings and subdued commotion. Another old man on the other side never stopped coughing and trying to leave his bed. He died the next night. At 5.30 every morning, we were assailed by the loud clatter of a trolley loaded with glass urinals being wheeled around. There seemed to be no escape from noise and disturbance. I was later transferred to the surgical ward and eventually had a thyroidectomy. I was very well looked after by the nursing staff and, when recuperating, the occupational therapist taught me to do tapestry, a satisfying and relaxing hobby which I took up again on retirement. Returning to work after two months was quite an ordeal. This experience as a hospital patient was humbling and I recommend it to all doctors! It made me more sympathetic and sensitive to those who were faced with the same challenge.

Family camp at Strathan Bay nr. Scourie 1931

THE MOUNTAIN ROAD TO SCOURIE FROM LOCHINVER.

The mountain road to Scourie 1931

Our 9 h.p. Wolsely car which could not get up the hill.1931

Cliffs at Handa Island Bird Sanctuary

*Ben MacDuhi (1309m.)with Pluto
1936*

Tedding hay on the farm 1938

4th Year medical students 1941

*Keir with his mother
on Graduation Day
1942*

Brother David in Royal
Signals, 7th Armoured Div.
1943

Staff at Western General Hospital 1943

Sister and staff nurse 1943

Keir and fellow house-surgeon. Christmas Day 1943

Nurses at Larbert Orthopaedic Hospital
1944

Taggart
1949

Groom David with Keir, Best man.
1950

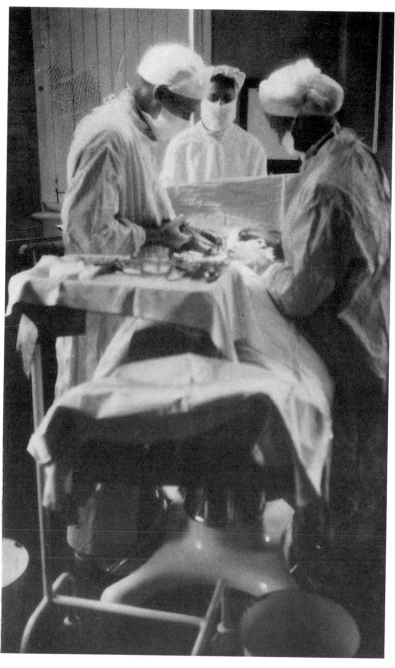

Operating at Willesborough Hospital, Ashford. 1947

Caledonian Ball, President and wife Betty. 1952

Receiving Order of St. John. 1986

Family picture. 1955

Attending patients at Whately Hall Hotel, Banbury.

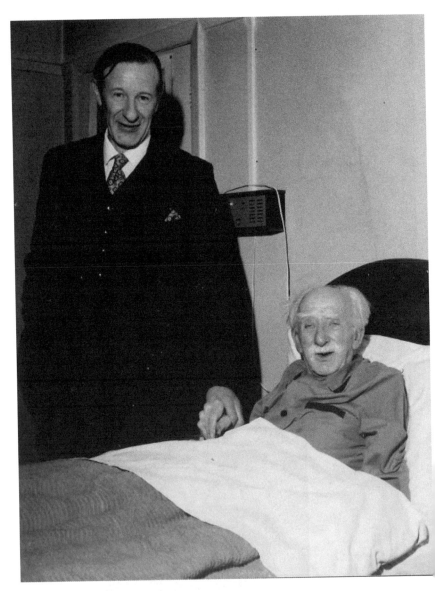

Attending Prof. Powell aged 92

Nigel and Fiona, Loch Lomond. 1960

Troopship Neuralia

Mother and Father, Winifred and Jimmy.
1964

Carden Farm, family home. Harvest 1964

Retired G.P. with catch. 1990

With fellow guns. 1990

Keir and wife Ann, Elgin Cathedral. 2003

Going fishing. 2004

Chapter 10: The Practice

As I gained increasing skills in my practice, I rapidly learnt that experience is a precious commodity. To be a good listener to the many and varied histories one heard from patients, rich and poor, and from all walks of life, is as important as conducting a careful clinical examination. At all times, the doctor must have lots of common sense and the ability to communicate effectively at all levels. I always found that every call and every new patient, each with a kaleidoscope of different problems, brought adventure and fascination rarely equalled in any other profession. General Practitioners bring new patients into this world, sometimes ease them out of it and, hopefully, at the end of a day's work, feel they have done a good job. Humility and mutual trust are central to a good doctor-patient relationship. It is essential that the medical profession can enjoy a well-earned reputation for providing care with sensitivity.

My closest friend in our Practice was an expert on children, having his D.C.H. (Diploma in Child Health). This was most useful, for if I had a difficult baby case I would ask Paul to give me the benefit of his advice. Similarly, if he had a surgical problem he would call upon me. Another partner and good friend, Gordon, also had a D.C.H., and Stephen with M.R.C.P. (Member of the Royal College of Physicians) had made a special study of diabetes and the interpretation of E.C.G.s. Also, Hugh, with M.R.C.G.P., was very well versed in General Practice before joining us. Caroline, who qualified in South Africa, was, besides being a skilled and sympathetic doctor, fluent in at least three languages. As a partnership, we encouraged full co-operation and the sharing of skills.

It was our aim to attend our own list of patients, even if partners would quite often meet one another in the same outlying village. We chose not to particularly zone our lists, for we felt that continuity in treatment was important. Thus we could get to know the families and their particular problems and, over the years, one increasingly appreciated the satisfaction of being in General Practice.

We were also blessed with Miss Moore (always called Miss M.), our irreplaceable private secretary. She was a great organiser, whose efficiency and know-how assured the smooth running of the business of our Practice.

One of the extra tasks practitioners were requested to do was to carry out medical examinations on applicants for life insurance policies. We examined clients for many insurance companies, such as General Accident and Norwich Union. This meant extra remuneration for our Practice. I very seldom had to fail an applicant, but sometimes advised the company to "load" the premium because of some medical problem like hypertension or diabetes. I must recount one amusing case I had. I asked this man if he would oblige me by producing a specimen of water. I gave him a flask and pointed him towards the toilet. He astonished me by returning almost immediately, and with a satisfied smile on his face saying, "Here you are, sir". He presented me with the flask now filled with water. In fact, just what I had asked for!

As a rule, my partners and I dressed well. I always wore a hat whilst on my rounds, at first a trilby, later a pork-pie. I became rather too attached to the latter and the Senior

Partner, a bachelor and rather particular, had a private word with my wife. He advised that my hat had become old and tatty and was no longer suitable wear for a doctor! I reluctantly bought a new one - this time a fore and aft. This caused my children, Nigel and Fiona, great amusement.

I shall never forget the Asian Influenza epidemic which reached this country in the early 1960s. The virus had spread from the Far East. It developed into quite an extraordinary and extensive epidemic, causing fatalities particularly in the young and the elderly. It lasted about three weeks, but intensively for two. The mainstay treatment was aspirin in some form and, occasionally, antibiotics. All General Practitioners were pressed, never had we worked so hard. I had never before driven my car so fast in built-up areas until visiting those afflicted with the virus. I always kept an eye on my rear view mirror, watching out for any police cars, but I got away with my speeding. Many police officers were off sick anyway! Our Senior Partner had an anti-'flu injection which caused a reaction and he was off work for two days. The rest of us decided not to have the injection and, fortunately, none caught the virus - we certainly didn't have the time to be ill! Our surgery staff were most helpful in giving advice over the telephone to a constant stream of callers; they often saved us doctors a visit. The smooth running of a doctor's surgery is so dependant upon a well trained team. At our surgery we were blessed with just that.

When calling at a patient's house during this emergency, I would sometimes knock at the door and be hailed from an upstairs window. The patient, saying that all inside were too ill to come downstairs, would throw the front door key down instead. Occasionally, I was asked to climb in through the kitchen window!

We had other epidemics over the years - of measles, chicken-pox and mumps, all before immunisation was introduced - but never on the same scale as that Asian 'flu epidemic during the early 1960s.

Betty and I continued the family tradition of camping during the holidays. Each summer, we visited my parents in Scotland and camped for two nights on the journey north and on the return. No motorways then to speed us along. I recall one occasion when the weather was so abysmally wet that we abandoned our site and sought out the nearest hotel at Bridge of Allen. This proved rather a disaster as Nigel, in customary high spirits, pushed Duke, our golden retriever, into the fishpond and the dog stank all the way home. Fiona had eaten too many chocolates and was sick in the car and Nigel had food-poisoning and the consequent diarrhoea. What a journey that was!

Another time we camped at an idyllic spot on the moor near Aberfeldy. Our first task was to stuff bracken into the sacks which were our emergency mattresses. Duke had just chewed our airbeds. We then, of course, dug a hole for the latrine to be screened by a sheet fixed to three poles. This was very hard work as the ground was so stony. Our tent pegs could mot be hammered in either and the side walls had to be held down with large stones. By early morning, the wind was so strong that our tents no longer gave us cover. Our spirits were raised when Nigel and I caught two trout in a nearby stream, which we cooked on a primus stove in the car boot. When we were ready to leave, one of the back wheels was firmly embedded in a boggy patch. We now had to dig the wheel free and jack the car up on stones. The children thought all this was a great adventure.

In later years, when I was financially better off, we went down to Newquay for holidays. There we all learnt surf riding at Fistral Bay. We became quite skilled and enjoyed this exhilarating sport.

The time came to leave our rambling surgery house in the centre of the town. I had bought some land on the outskirts of the delightful village of Bloxham, just 4 miles away. I commissioned a reputable architect to design a large house to comfortably accommodate our family, including my mother-in-law who had been living with us for some years, since her husband's death. A local builder was employed to build the house and we moved in during the spring. Further along the road was a huge, permanent caravan site, well set back from view. Many caravans were owned by travellers of Romany origin. One winter's night, about 1a.m., a loud knocking at the front door woke us all up. The caller asked me to come straight away to see an old lady taken ill in one of the caravans. I said I would just slip on some warm clothes and walk the half mile to the site, but was immediately told there was a car waiting to take me. Indeed, there on my driveway was a beautiful Rolls Royce! I attended the patient who had suffered a mild heart attack. The family asked me my fee - I said a fiver would do and the son took a huge wad of £20 notes out of his hip pocket and gave one to me, insisting he did not want any change. I was then taken home, this time in a large Rover car. Well, I visited the old lady a number of times and she always refused to become a National Health patient. I also attended others on the site, especially children who had developed the usual childhood infectious diseases. Some of the caravans were very luxuriously furnished and others, of course, were more basic. I would like to add that I was always treated with extreme courtesy by the Romany people who lived in them.

Throughout my practice career, I attended many patients of all ages and from all levels of society. With very few exceptions, I enjoyed meeting, conversing with and treating each and every one of them. For thirty years, I looked after all the nuns at the local Priory as well as the few priests living in the Presbytery in the town. I was not a Roman Catholic, having been brought up as a Scottish Episcopalian, however, I did have a great respect for these dedicated and selfless workers. The nuns, mostly ex-head teachers, ran an excellent school for children up to the 11+ examination age. The priest in charge of St. John's R.C. Church was a delightful character, later to become a Monseigneur, in recognition of his services to the community. For many years, he organised pilgrimages to Lourdes and to the Holy Land. After my retirement, my wife and I joined three of his tours to Israel, each turning out to be a rewarding and exciting experience. The last one was following in the footsteps of St. Paul and this was particularly interesting for us as we journeyed from Israel to Greece and finally to Ephesus in Turkey.

The Monseigneur became a valued and true friend of mine. He was especially helpful in dealing with some of my patients who had suffered a bereavement or who were terminally ill or perhaps under extreme stress. I admired his energy and his commitment to work - sometimes at the expense of his own health. The physician heals the body but the minister heals the soul.

As my seniority in General Practice became established, I was required to attend medical meetings which were held to discuss local health issues and general N.H.S. policies. I took my turn in acting as Chairman of the North Oxfordshire Group of Doctors and was a representative on the local Medical Committee and Area Health Authority in Oxford. I was a reluctant committee man, preferring to do the job I was trained for - being a doctor. Others, however, were more suited to operate in the arena of medical politics and they are essential and indispensable members of our profession.

Chapter 11: More House Visits

I had many and diverse experiences, as all doctors have in their working life. But, permit me to recount some that I remember most vividly. One of the first patients I visited was a very prim and proper elderly Scottish lady. She presented with abdominal pain which turned out to be diverticulitis - inflammation of the large bowel. As I approached her bed, the toe of my shoe hit the chamber pot stowed underneath, giving a loud, dull clonking sound. I apologised and looked under the bed for fear that I had cracked the pot, only to find that it was full of water. I asked the lady about this discovery and she replied, with great enthusiasm, that she always kept her bed chamber pot full of water as this helped to keep her rheumatism under control!

Another patient I well remember was an elderly lady, probably of Romany origin, who lived for ten years on nothing but Guinness, white bread and cheese. She had no teeth and refused dentures. She lived with her daughter and her family and died at the ripe old age of 87 years. At one time, I wrote an article in some medical magazine suggesting that a daily glass of Guinness for nursing mothers helped to stimulate lactation. The Guinness Company sent me a dozen bottles of their product and I, in turn, presented them to this old lady. She was more than delighted and shared the odd bottle with me.

I frequently attended elderly patients living in a collection of council house bungalows on a small estate. This was remarkably remote from any shops and also on quite a steep incline - a very short-sighted housing policy on the part of the Housing Department of the Town Council. One night, about 2a.m., I had a call from an old man who said that his wife had fallen out of bed and he could not get her back again. They did not enjoy the luxury of central heating and the weather was very cold and, to avoid hypothermia, the sooner she was put back into bed, the better. I knew this couple well. Sure enough, when I arrived, I found this overweight old lady, who had a vast posterior, on the bedroom floor. Fortunately, I was pretty strong, but it did take a lot of effort for both myself and the husband to get her, first of all, on to a chair. This was quite a complicated task involving much manoeuvring and accompanied by many grunts and groans! We first placed the chair on its side and managed to roll her onto it. We then heaved the chair, complete now with the old lady, onto its back and then heaved again, finally upending it so that the patient was, at last, sitting upright. She was most co-operative but not exactly helpful. We then positioned the chair up against the side of the bed and managed to literally roll her on to it. All three of us ended up having a whisky together - well-earned!

I can remember having to make frequent visits to a patient who lived in a bungalow two doors down from last couple. She was a sour and selfish old woman who led her poor husband a permanent "merry dance". She was incredibly demanding of him. He often wrote me letters asking me to call, and always headed the letter, "Good morning, Doctor". One day, I had to attend this lady – Fanny, was her name - because she was very poorly with a bad cough and high temperature. I needed to examine the base of her lungs at her back, which would involve her nightdress being pulled up - but she flatly refused to allow this. I asked her husband for his help. The only way for the two of us to pull up her nightdress, which consisted of yards and yards of long material, was to lift her up a little, as best we could. We did this on the count of three and, on "Three", I said, "Heave!" Well, with a huge tug at her nightdress, we only managed to split it all the way up to her neck, causing a loud tearing sound! Fanny was furious and she turned and slapped her husband's face saying, "You stupid bugger - you've torn my nightie!!" I gave her poor husband a wink and now managed to examine the base of her lungs. I found both lungs had some degree of pneumonic consolidation and advised that Fanny should be admitted to hospital forthwith. This advice she flatly refused to accept and so I had no choice but to treat her within her own home. It was an uphill struggle to get Fanny better and she presented a considerable burden and responsibility for her husband. However, with the help of a wonderful District Nurse and penicillin injections, we won the day and the patient was restored to good health. Her hen-pecked husband waited on her hand and foot and attended every beck and call. Two years later, this uncomplaining man sadly suffered a severe stroke, from which he died three days later. Fanny then had to manage for herself. It was quite extraordinary and surprising how well and courageously she coped on her own. She could not go outside, but she was helped by Social Services, Meals-on-Wheels and the District Nurse, plus regular visits from myself. She lived on for another two years.

    Obstetrics formed an important part of our Practice. With the help of midwives, we attended confinements both on district and in hospital. In all humility, I never failed to feel joy at the first cry of a newborn baby. Sometimes when the labour was prolonged, I felt afraid that something could go wrong. However, as long as the mother and baby's conditions were satisfactory, I learnt that "masterly inactivity" was best, as taught by Professor Johnston - professor of obstetrics at Edinburgh University. I admit it was sometimes difficult to decide when to refer a case over to the care of a Consultant Obstetrician. Furthermore, that decision was best made early in the confinement.

Here I would like to recount how it came about that I shared a bed with the District Nurse! Her husband was the Senior County Councillor. She also happened to be an excellent midwife and helped to deliver some patients of mine at home. One confinement, as usual in the middle of the night, took longer than expected, for the labour was prolonged. It was going to be a marathon wait. The husband, who had recently been released from prison, was most concerned and suggested that the nurse and I may like to rest by lying down on the bed in the next room. This we did for about an hour, alongside their three young children! Unconventional perhaps, but nevertheless a welcome respite. The baby eventually arrived and we all celebrated with a glass of beer.

In any difficult confinement where forceps or instruments were required, a local anaesthetic was all that was necessary. The forceps used were firstly sterilised by placing them in a pan of boiling water, with the handles sticking up out of the pan, of course!

Another birth I had to deal with, and without the help of the midwife, was one afternoon during a snowstorm. I was called to a gypsy caravan, which had no separate compartments and just a dome shaped roof. The caravan was in the centre of a scrap yard. The mother-to-be was just about to give birth to her firstborn when I arrived. I quickly washed my hands with pure Dettol and delivered a baby boy. The placenta, or afterbirth, quite quickly appeared afterwards, I administered an intramuscular injection of ergometrine to induce the uterus to contract and to stop the bleeding. All was well. I then put both mother and baby into my car and took them to the maternity wing of the local hospital. The baby was later christened with my name, which I felt was a great honour. The father had been a German P.O.W. He was a hard worker, a labourer, and they were allocated a council house, where his wife had three more children, all delivered at home. I still get Christmas cards from this family, giving me their latest news.

One of the dreaded complications of obstetrics in general practice is the diagnosis and treatment of placenta praevia, either total or partial, where the placenta, or afterbirth, lies in front of the baby's head, thus blocking the baby's birth. If I was suspicious of this on ante-natal examination, I immediately referred the patient to the hospital Obstetric Out-patient Clinic or even admitted her to hospital as an emergency. But two cases did give me a few grey hairs! The first was in my consulting room. A patient of mine who had been attending regularly for antenatal observation was now about 32 weeks pregnant. This was her second pregnancy and she had had no problems with the first. When I examined her abdomen, I found that the baby's head was still not engaged in the pelvis - nothing unusual at that stage. But I could not push the baby's head down. I thought this warranted a vaginal examination as I was suspicious that the placenta was lying low. As I proceeded with this, a little blood appeared and then suddenly a huge gush of blood. I quickly towelled her up, and sent immediately for the ambulance. I explained what had happened and she remained remarkably calm, accepting the situation. The bleeding had stopped by the time she was admitted to hospital but she did have a blood transfusion later. The patient was sedated and the Consultant decided to leave well alone. She was kept in the hospital and was eventually delivered of a slightly premature, but healthy, baby two weeks later. She had a marginal placenta praevia, that is, the afterbirth was only slightly in the pelvis.

The second difficult case was a lady who lived in a village about five miles out of town. She was around 30 weeks pregnant for the second time, had had no difficulty with her first baby, and I had seen her in the surgery for routine follow up two weeks earlier. I received a call in the early hours of the morning from her husband, who said that his wife had abdominal pain and that she had lost some blood. When I arrived at the house - after a pretty fast drive - I was staggered to find the patient bleeding profusely. There was a lot of blood on the bed. I immediately sent for the emergency blood transfusion team which had only been organised six months beforehand. They were quite wonderful and highly efficient. I also sent for the ambulance. I gave the patient a sedative injection and made a note of this for the hospital. She had a Caesarean section shortly after being admitted to hospital and her baby was safely delivered. The mother and baby convalesced well and they were discharged, very happy, ten days later. I was a very relieved doctor!

The Consultant Obstetrician, who successfully performed this operation, was a good friend of mine. Occasionally, he asked me to assist him in Theatre, which was indeed a great pleasure. Not only was he a masterful surgeon but also an expert and highly skilled rock climber. He had once climbed the North face of the Eiger with two companions and had become marooned during the ascent because of sudden severe weather. The men could neither advance nor retreat for 24 hours. One died of exposure, but my friend managed to survive by wrapping himself in a double-skinned polythene cover.

In my profession, one is trained to view the naked body with complete detachment. A GP on his daily round never knows what will be facing him. I have had my full share of unexpected encounters. One early morning, well before breakfast time, I was called out to see a child complaining of abdominal pain and who had been vomiting. I knocked at the front door and, as usual, went straight in saying, "Good morning". There was no reply and nobody about. So I went upstairs and again said, "Good morning" but still had no reply. I knocked at the first bedroom door, and went in to find the room empty. Then I knocked and entered the second bedroom and to my complete surprise found a totally naked young woman on the bed. We were both taken aback. She said, "Oh dear!" and tried to cover herself. I said, "Please don't worry. I'm not a man, I'm only a doctor!" Apparently, the mother of the ill child had gone to work and had left her daughter in charge. I admitted the child to hospital with a suspected appendicitis.

Another time, I had to deal with Connie. She had been a resident housemaid to two elderly bachelor brothers for many years. She eventually had to leave as one brother died and the other had to be admitted to a nursing home. Connie was allocated a small council house. She was a diabetic and often did not give herself the prescribed dosage by injection, which was essential to keep her blood sugar at a safe level. Unfortunately, she developed gangrene of one foot and had to have her leg amputated. This is one of the unhappy complications of diabetes. One morning, I was called to her council house by a neighbour because Connie had not taken in her milk bottle; it was still on the front doorstep - always a sure sign that something was amiss. The neighbour said that Connie did not open the door when the milkman rang. When I arrived, I could not get into the house so I borrowed a ladder and called the Police as I did not think I should invade Connie's house without the "law" on my side. The Police officer duly arrived and I went up the ladder and managed to climb through an upstairs window which was slightly ajar. I called Connie's name and told her who I was. I found her in the bath where she had been all night, unable to lift herself out. She had cleverly kept herself fairly warm by trickling hot water into the bath. It was in the early spring so the house temperature was quite cold. I managed to lift Connie out, dried her with a towel and covered her with as many warm clothes as I could find. Meantime, the ambulance arrived and she was treated for hypothermia and shock. I'm glad to say that Connie recovered and soon afterwards was given accommodation in a complex of warden-operated flats.

On another early morning visit, I was again met by the unexpected. I was called to a house in a village to see a teenager who had been sick all night. I knocked at the door and, as usual, walked in saying, "Good morning". A young woman appeared, dressed in a very short nightie. I didn't think to be embarrassed - doctors just don't - but she said in a surprised and somewhat annoyed voice, "Who are you?" I said I was the doctor and I had been sent for to see an ill boy. "Oh," she said, with relief, "that will be next door." I apologised profusely and went next door, as I should have done originally. The lady I met turned out to be the wife of a top referee in professional golf matches, and by a strange coincidence I was introduced to both husband and wife during a social function a few weeks later. However, this time the introduction was more formal and we joked about my uninvited early morning visit! There were many more similar such incidents; being a GP, these simply cannot be avoided.

In large council housing estates, it was sometimes very difficult to find the numbers of the houses, and often houses were tucked away in little alleyways. I had a call to see a child at about 11p.m. one evening. It was dark and just beginning to rain. I parked my car in a car park in the vicinity and was wandering about with my torch and bag looking for the number of this house. An obviously rather inebriated man appeared, asked my problem and very happily insisted upon taking me to the address. He put his arms around my shoulders and we both rather staggered to the house which turned out to be his own home!!! We were both laughing about this but, when his wife came to the door, she swore at her husband and gave him a good push. He fell over backwards into the front garden, I was invited in, not daring to turn round to see if her husband was all right, and proceeded to deal with the ill child. When I came out, the husband was full of apologies. I thanked him for helping me and went on my way home, relieved I had no extra patient to attend.

A young, pleasant looking woman came to my surgery feeling very depressed. I had delivered two of her children some years before. She told me her husband had gone off with someone else and had taken the children with him. I gave her a prescription for about twenty librium or valium tablets and asked her to return in a week's time. About four days later, one of her neighbours telephoned me to say that the milk bottles had not been taken in. When I arrived at her council house, all the curtains were closed and the doors locked. However, with the help of the neighbours, I managed to open a window at the back of the house and climbed in. I found this poor woman unclothed, lying deeply unconscious on the sitting room floor. I shouted for the neighbours to summon an ambulance immediately and to notify the hospital. Before the paramedics arrived, I managed to find some clothes and covered the patient as best I could. Later, I went to see her in the hospital, where I found her recovered and full of apologies for the drama she had caused. I'm glad to say that she found happiness with another partner and, shortly afterwards, left the area. Bless those milk bottles!

Chapter 12: Consultations

One evening, when I was in my consulting room, I suddenly heard a loud shout of "Help!" from one of my partners and some loud screaming. All hell was let loose in the surgery. Together with two other partners, I rushed to see what was happening. We found this large woman, screaming her head off, and struggling against the doctor who, by this time, was sitting on her chest and trying to restrain her. She was a known mentally unstable character and had just gone dangerously berserk. With great difficulty, we gave her a strong sedative injection and sent for the ambulance to take her straight to the mental hospital. I telephoned the psychiatrist who told me he could not admit the patient without first doing a domiciliary visit for which, of course, he would get well paid. I was not at all sympathetic to this suggestion and told him bluntly that the patient was on the way.

I rather think that in my group practice, I had more patients registered with me from Scotland than my partners. Two families certainly gave their medical cards to me. They came from Peterhead on the north east coast and all had the very marked Aberdonian accent peculiar to that part of the country and fishing community. They really spoke the "broad Doric" and, being a fellow Scot, I was the only one who could understand their rustic dialect! It was my pleasure to be their medical practitioner because they were a delightful and hard working family. They had a fish and chip shop and a general store on a housing estate.

I was presented with an amusing and embarrassing problem by a young, attractive lady. She told me she wanted to get rid of her boyfriend, and asked for my help. Well, she did, in fact, suffer particularly from a recurrent and nasty vaginal discharge. This was treated by prescribing the usual vaginal pessaries, but in a week's time the copious and smelly discharge recurred. I thought that the etiology, or causation, of this recurrence lay firmly on the shoulders of the boyfriend and I agreed to help the victim, my patient. I asked her to co-operate with an idea I had in mind and this was to be quite confidential. I told her how to paint herself internally with Gentian Violet. This was the old-fashioned way of treating the condition, before the more modern medicated pessary was available. She readily agreed to do this. Well, sure enough, the boyfriend arrived and had sex with this woman, and again in rather a cruel way, as was his custom. When he had finished, he was horrified to find that his private parts were covered with the startling purple dye. He dressed hurriedly and left the house in a furious panic - never to return. The gentian violet had done the trick! The lady in question was extremely grateful and later presented me with a bottle of wine. I was delighted with the result and further delighted when she visited my surgery some months later to say that she had met a new and decent man and that they were leaving the town, hoping to get married in the near future.

Over a six month period, I had a tramp who visited my surgery once every three weeks to have his varicose leg ulcer dressed. His life was taken up by simply walking from one village to another and back to the town. He wore gum-boots which he never changed. He slept rough and had reasonably good warm clothes and a few other possessions which he carried in a sack over his shoulders. He was a man of about 35 years of age, wore a cap and sometimes was bearded, other times shaven. I dreaded his coming in as we all did, because his leg absolutely stank! But he was a patient and had to be treated. I felt sorry for our practice nurse when she and I had to remove his gum-boots and dress his leg ulcer. Every time he visited, we nearly emptied an atomizer air freshener can all around the surgery! This situation was solved by an extraordinary coincidence. An older woman, living by herself, took pity on the tramp and gave him a room and bed in her house. In exchange for this, he tended her garden, her rabbits and her chickens. It was amazing how this chap's life changed for the better so suddenly. He found a new purpose in life and became clean, considerate and cordial. The consulting room presents so many varied problems and variety is the spice of life and, for that, I loved my job as a doctor.

I had one patient who occasionally wrote a poem to me after I had dealt successfully with some malady or other. I found this quite touching. Whilst on the theme of notes from patients, permit me to quote a few. Some were sent to me and others to my fellow practitioners. They begin "Dear Doctor":-

"Could I please have some suppositories to fall back on?"

"I have heard a lot about you. I know you are tops on children but how are you on women?"

"My fiancée and I are both sterile. Is there any danger of passing this on to our children?"

A reply to a Consultant when doing his ward round: "Yes, thank you sir, I feel a lot better since the young doctor passed a cathedral this morning."

To a Consultant from a doctor: "I would be grateful if you would arrange for Mrs A. to be confined to hospital as she is an elderly primate."

Chapter 13: Police Surgeon

Shortly after I became a partner in the group practice, I was appointed Police Surgeon of the area. The Senior Partner had been the Police Surgeon for 28 years and he asked me to follow him. This was quite an honour, particularly as it was so early in my career. I much enjoyed this part of General Practice and, in fact, remained the Police Surgeon for thirty-two years. I found it immensely satisfying to take an active part in other fields of medicine besides just being a General Practitioner.

I made many new friends in the Police force and always found my Police colleagues very helpful. As an undergraduate, I had always been fascinated by Forensic Medicine. The then Dean of the Edinburgh Medical School, Sir Sidney Smith, was the Professor of Forensic Science. He proved many times that Forensic Science can be harnessed to solve many baffling crimes.

It was in 1952 that I was appointed Police Surgeon. Many of the problems I was called to deal with were "drunk-in-charge" cases, mostly men but occasionally women. They would be brought into the Police Station by the arresting Police Officer, who would be suspicious that the accused was drunk in charge of a car. The majority of these calls were in the evening and usually before midnight. In those days, a complete and careful clinical examination had to be made and meticulous notes written. I quickly learnt that if Court proceedings were instituted, the defending barrister would put many searching questions to the medical examiner - and that was me - to try and trip him up and make his medical evidence inadmissible. A clever barrister could so easily pick holes in the medical report. Standing in the dock in Court, which I often did as the Police Surgeon and the Expert Witness, was sometimes a daunting experience. One of my first drunk-in-charge cases was a motorist who was very drunk indeed. He failed all of the tests I put to him. He was found guilty of being under the influence of alcohol to such an extent as to be incapable of having proper control of a motor vehicle. No one had been convicted of this particular offence in the area for eleven years! I became an expert on alcoholic intoxication and the physical and mental effects thereof and, on the strength of this, I was asked to lecture Police Officers at Thames Valley Headquarters.

The clinical examination of a suspected drunk-in-charge was a lengthy business. Apart from the routine of taking the blood pressure, temperature, examining the heart and lungs, abdomen and all tendon reflexes, I asked the accused to repeat "The Leith police dismisseth us'' or "British constitution" three times. He then had to walk along a chalked or taped straight line and furthermore was asked to stand on one leg for 5-6 seconds in the Highland fling position. I always carried out this particular test along with the accused. I may add this always caused some hilarity among my Police colleagues! After this rigmarole, a specimen of urine was taken and this would be sent for analysis to the nearest forensic laboratory.

One inebriated man, when asked to quote "The Leith Police dismisseth us" replied somewhat plaintively, "But I don't come from Leith, Doctor, I come from Banbury." And when I required him to walk along the white tape he commented,

"But that's not straight!" At the end of the clinical tests, he was duly charged with being under the influence of alcohol.

Another suspect I had to examine was a very large and menacing gypsy. He adamantly refused to cooperate. When I asked him, very politely, to walk along the taped line, he came up close, held his fist at my face and shouted, "Not on your Nellie!" I asked him to provide a specimen of urine and again he shouted, "Not on your Nellie!" I told the Sergeant on duty to be patient. The accused flatly refused to sit down but wandered about the room which had a wooden block floor. I knew he had had far too much to drink and after some time, a trickle of urine appeared on the floor from the gypsy's trousers. I told the sergeant to take a sample from the floor as our required specimen! Next morning, as happened in most cases, the gypsy was full of remorse and entirely polite. In later years, of course, the breathalyser was introduced, where the suspect was requested to blow into a tube which registered a degree of alcoholic intoxication. I was then often called to the Police Station to take a sample of blood for analysis. This whole new procedure was so much quicker and more accurate than the previous clinical examination and random tests. All Police Surgeons were delighted by its introduction.

Many of the Police requests I answered were to take a sample of blood from a victim of assault or from a suspected criminal and also to inspect wounds, bruises and other bodily injuries. I had to give an opinion as to the cause of these injuries, the time they occurred and the duration of healing. When giving evidence in court as an Expert Witness, one had to have made very careful measurement of wounds and full notes of all injuries sustained.

Sometimes, I had to examine suspected culprits in order to decide on medical grounds if they were fit to be detained. One morning, I remember being called to Court and found I had to examine eight prisoners. This occurred during a Prison Officers' strike. When I arrived, I was welcomed by the Police Inspector in charge and ushered into the room where the prisoners were being held. The door was shut and locked behind me, much to my alarm. But the Police Officer did reassure me that he would be standing close to the door on the other side and I was to shout if there was any trouble. Well, I proceeded to examine the chest, heart and ears of each man. I also smiled and was polite, sometimes expressing regret that I had to do such a duty. I also added that if I found any man ill, I would report him as unfit to be detained in a prison. One older man, whom I thought had been in "clink" before and thus knew the ropes, said he suffered from breathlessness and had pain in his heart. His heart sounded normal to me and I asked him if he was on tablets and, if so, which ones, and the exact dosage. He said he was on digitalis but the dose he told me he was taking would have killed him! I passed him, but I mentioned I would make a sympathetic report on his behalf. I left that room somewhat relieved. You know, a smile and a polite approach often disarms the belligerent.

One unusual Police call was to certify the death of two men involved in a house fire. This was quite late at night and, as usual, I put my "call out" clothes on top of my pyjamas. When I arrived at the scene, the fire engine had put out the fire but, of course, there was water everywhere on the ground floor. The bodies were upstairs. I still had on my bedroom slippers and did not relish paddling and getting my feet wet. This predicament was very obligingly dealt with by the Chief Superintendent of Police, who carried me on his back across the floor to the staircase! Both victims were very severely burnt and charred. The smell of burnt flesh was truly awful. After a brief inspection of the bodies I was pig-a-backed again and I returned to my car. This obliging Superintendent later became Chief Constable of the Isle of Man.

I attended a number of road accidents to certify a death or to help a trapped accident victim. Some of these experiences were upsetting and frightening and made me drive more carefully. The worst accident I dealt with was particularly perturbing. There was a huge traffic jam and I had a Police escort to the scene. I put on my headlights and flashing hazard lights and closely followed the Police at speed, passing many stationary cars. The scene I came upon was chaotic. There was a wrecked car up against a tree and bodies were all over the road. I certified the death of four young soldiers and the driver of the car. Nobody else was involved. The driver had apparently picked up these soldiers who were on their way home on leave. The road was very twisty and there had been accidents there before. The driver was thought to have been driving too fast thus failing to negotiate a corner and crashing into the tree. This stretch of road has since been made a dual carriageway for about a mile.

I attended a number of other fatal accidents but another I do remember, vividly, for this was to certify the death of a very attractive dark haired girl early one morning, about 7a.m. She had failed to control her car on a nasty corner on this "B" road, and had crashed into a tree. Her neck was broken. Since then, whenever I passed that tree I always saluted - a bit emotional, but there it is.

About two o'clock one morning, I had to help at an accident where a fully loaded lorry had left the main road on a bend and had overturned. The driver was trapped underneath and was shouting for help, obviously in a lot of pain. On all night calls, I always wore an old set of clothes which I donned over my pyjamas. On this particular night, I also had on my gumboots and leather gloves. This was fortunate because of the mud and broken glass at the scene. I decided to give this man an injection of pethidine as soon as was possible. We were waiting for the fire service and more help to arrive to prise the lorry off the casualty. I had to crawl some distance under the vehicle and a Police Officer held on to my ankles. I tried to comfort the driver, gave him the injection and stayed with him until it took effect. There was hardly any space to wriggle backwards from under the lorry so the officer tugged vigorously at my clothes and gradually pulled me out. By this time, my trousers and pyjamas were around my ankles and he humorously remarked, "Ah, doctor, I see you're a man after all!" The upshot of this accident was that the lorry driver recovered in hospital in spite of some severe injuries.

One other accident was quite spectacular. This time, the motorist was trapped in his car, which was lying on its side. When I arrived, I was surprised to learn that the driver was a doctor from some distance away and had been motoring to see a relative. The ambulance crew and Police Officers helped me to release him and I used my thick crowbar, which I always carried in the boot of my car. The doctor had obviously fractured his femur and I gave him an injection of pethidine and, as usual, wrote the dosage on his forehead to alert the hospital.

One of the most horrifying calls I had was to certify the death of a man whose parachute had failed to open. He was one of a large group of men giving a display of parachute jumping. He was, of course, very dead, and his body was just a flat replica of a human being. He must have fallen from a great height, for there was quite a hole in the ground where he had first landed before bouncing just a few feet away. I should think every bone in his body must have been shattered.

Suicides were not a common occurrence. However, I did attend some cases of drowning, mostly in the local canal but sometimes in ponds at a remote location. It was often difficult to estimate when death by drowning had occurred. In any case of apparent suicide, one always had to keep an open mind and examine the surroundings and the body, especially the neck, for any sign of violence. The eminent Forensic Pathologist, the late Sir Keith Simpson, said, "The Police Surgeon is the watchdog of the public, and must keep an ever open eye for the kinds of death that require an explanation." I saw a few cases of death by gassing, suicidal hanging and some by gunshot wounds. All these fatalities were harrowing, but the last mentioned were particularly awful to view. One had to be just so sorry for the distress caused to the relatives. One example of attempted suicide was a man who had cut his throat, late one night. The neighbours had called me. There was blood all over the man's clothes and on his bed. He said he was right handed when I asked him. Sure enough, high up on the left side of his neck were several small tentative cuts, before the big wounds. This, then, was the textbook forensic sign of attempted suicide. He was admitted to the hospital and I was on duty that night for emergencies. I stitched him up with a good many stitches. Fortunately, nothing major had been cut. I ticked him off for being so stupid, especially as I had to get out of bed to treat him! He was not registered with a doctor but later came on my list and was grateful for my help and advice. He quickly settled down and got a job and then left the area for better paid work.

Quite a number of my Police calls were to certify the death of someone in their own home, having been found dead by a relative, neighbour or the Police. In these cases, as a Police Surgeon, one's duty was to be unbiased as to the cause of death. One was careful to observe the surroundings in the room for any sign of a struggle, for a bottle containing sleeping tablets, for any note left by the dead person and finally for any sign of injury to the body. The Police Officer in charge likewise made observations. The usual conclusion was death by natural causes. One extraordinary suicidal case was where a man had thrown himself in front of a passenger train some distance up line from the station. The driver, who was understandably very traumatised, reported the incident to the police. When I arrived at the scene with three Police Officers, it was quite extraordinary to find bits of body here and there on the railway track extending over a radius of at least 200 yards. One of the police officers was sick on the spot when I identified one gruesome mess as half of the man's chest and part of the neck, no heart - that was discovered further down the line.

One unusual case was of a lone hiker who had decided to follow a very old walking route he had found on a map. He had been reported as missing, and, after days of searching, he was eventually found down a deep well. This was close to a hedge and the unfortunate fellow had fallen down the hidden shaft. The Police called me to certify his death and he was winched up from the well, which only had a little water in the bottom. I concluded that in falling down he had very quickly died of a broken neck.

One of the many tasks given to the Police Surgeon was that of dealing with sexual offences. This often was the rather sordid side of Police work. However, it was my job and I had to get on with it. It was always very difficult to prove a case of rape as it was quite unusual to find the required evidence of severe trauma in the woman examined. Unlawful sexual intercourse or sexual abuse or harassment was likewise difficult to prove on forensic investigation. Swabs taken at the examination were crucial and also specimens of hair and clothing for evidence of semen. Any finding of this would convict the assailant. I did have one chilling case of a young woman who accused her neighbour, who was a professional and well respected local figure, of having sexual intercourse with her against her will. She was a very angry, upset and distressed lady. They lived in a town some 15 miles away and every doctor there declined to examine this case. Physically, I found some very obvious bruising of the woman's perineal area and of her thighs. Vaginal examination was painful for her. I took swabs, of course, and specimens of hair and clothing, each and all carefully labelled as one did in all of these types of case. I then suggested that the neighbour, who persistently denied rape, should also be examined. The Police were not very keen on this idea, as a car had to be sent to ferry the man from the distant town. I could not find anything in particular, but I did take swabs from his penis. I was delighted, as were also the police, that the forensic laboratory reported the finding of vaginal cells on the swab which matched those taken from the woman involved. This proved the case of rape and he was charged with this offence in Court.

I had to deal with three murder cases. It was obvious that murder had been committed, but to prove this in Court, very thorough investigations had to be implemented. The first case was that of a girl aged about 18 who was found at the side of a field near the main road. She had been given a lift in a taxi and the taxi driver was later accused of her murder. When I found her, she was lying partly on her face and partially clothed. I did not move the body, but took careful notes of any sign of a struggle i.e. flattened grass here and there - there was none. So I concluded that she had been dumped there when she was already dead. I examined the girl for signs of rigor mortis for this helps in estimating the time of death. I took her axillary and mouth temperatures and also the temperature of the air and the ground underneath. The police photographer had already taken his full quota of photographs as was always done in any assault case. The body was then taken to the hospital mortuary where a Home

Office pathologist carried out a post-mortem. One shudders to think just what torture this poor girl had endured.

My second case was the murder of a patient of mine. She was a cheerful and pleasant woman of about 28. She had been married, and I knew she currently had a succession of partners. One evening about 8p.m., I was motoring home after my evening surgery, when I noticed her standing at the roadside trying to thumb a lift. She did not recognise me, as it was fairly dark, but I noticed that she was all "dolled up" and wearing an ultra mini-skirt. I motored on, wondering to myself what man she was to be with this time? The following day the police called me to her council house. I was shaken to find her lying naked on her bed in a pool of blood. She had been stabbed at least six times and was dead. I made notes of what I found, certified her death and the Home Office pathologist was called. The murderer was later found and charged.

My third murder case was an interesting case of forensic observations. I had to visit a camping site where a caravan was badly damaged by fire. Lying on the ground nearby was a fairly stout, partially dressed and very dishevelled woman of about 45 years of age. I admired the ambulance paramedics, for they were attempting mouth-to-mouth artificial respiration. Her face was very suffused and cyanotic. I concluded she was dead, as I could not detect any pulse nor hear any heart sounds with my stethoscope. On examining her body, I just fortunately noticed some very suspicious finger nail marks on her throat. It appeared to me that she might have been strangled in the caravan which was probably set on fire to disguise her death. I reported my suspicions to the police inspector in charge and suggested that the body should have a post-mortem by a Home Office pathologist. I was then asked to have a look at a man in another caravan nearby who was reported as having a badly cut throat. I found him lying in bed with his clothes very dishevelled and with a large wound in his throat, which was not bleeding. He said he was the dead woman's partner and when the caravan had caught fire, he had pushed her out and managed to escape himself, but in the process had cut his throat on broken glass. Well, I looked at his throat very carefully and then asked him if he was right handed - the answer was in the affirmative. Sure enough, I spotted several tentative small cuts high up on the left side of his neck, and then further round was this large gash. This was not particularly deep and no major blood vessel had been injured. I informed the police inspector that I thought this was a case of attempted suicide. I did not tell the injured man this nor anybody else. He was soon taken to the hospital, and there his neck wound was stitched up and dressed. He was later arrested and charged with the murder of his partner.

The distinguished Forensic Scientist, Professor Camps, once wrote on the investigation of crime: "Fools and wise men are equally harmless. Dangerous are those who are half-foolish and half-wise and who only see half of everything."

About a year before I retired, the Police involved me in a most exciting and somewhat frightening situation. I was called to attend and help a man who had fallen down a 60 foot high silo used to store grain during harvest time. Two silos had already been built and the third one, into which the man had fallen, was nearing completion. When I arrived, I was told that the only way to reach him was to climb up a vertical ladder inside the first silo. The constable confessed that he was glad he did not have to do this because he disliked heights. So I started to climb up the ladder, holding my doctor's bag with one hand and the ladder rail with the other. It was quite hard work - 60 feet is a long way - and I dared not look down! When I finally reached the top, I had to walk across the middle silo on a rather narrow wooden plank. There was a sagging rope on one side which gave very little support - not very comforting. I took a long time to cross over that silo, all the time looking straight ahead. The plank swayed a bit as I walked and I must admit I was so nervous my pulse rate must have doubled. When I thankfully reached the platform of the third silo, there were two other men waiting for me. The casualty, who was obviously in a lot of pain, was lying at the bottom of the silo, accompanied by another man who had been lowered down earlier. I loaded a syringe with pethidine, a rope was tied around my chest and shoulders and I gradually descended. It seemed to take forever. Finally reaching the bottom, I found that the poor fellow had a compound fracture dislocation of one ankle. A stretcher was then hoisted up from the outside of the silo and eventually down to us. The other helper and I strapped the injured man securely to the stretcher and he was hauled up to the platform, followed in turn by the two of us. Finally, the traumatised casualty was lowered extremely slowly and carefully over the side of the silo. It was a long and most harrowing operation and we were all extremely relieved when it was over and the victim had been whisked off to hospital. When I got back onto firm ground, the Police Officer congratulated me, but I told him it was all in the day's work. I was a very frightened doctor at one stage, but it really was an exciting experience. When I returned home before my surgery, I had a good dram of whisky! This was against my rule as patients might smell my alcoholic breath - but it was worth it.

In the early 1960s, a night-club opened in Banbury town. Here, a band or disco would provide dance music from 9p.m. to 2a.m. the following day. It was hailed as a great innovation and intended to provide amusement and a means of socialising for the younger generation. After a promising start, unfortunately and sadly, this night-club fell into disrepute.

About this time the Police Authorities were becoming aware that stimulant chemical compounds were being misused and sold as drugs. These belonged to the Amphetamine group and normally were prescribed by doctors as an anti-obesity treatment or to counter stress and depression. They had a sympathomimetic action and when taken in large doses acted as a stimulant, giving a feeling of well-being, increasing energy and enhancing sexual desires. These Amphetamine compounds were prescribed as Drinamyl or Durophet, but on the drug market were known as French blues, purple hearts and black bombers. It became more common for young people to take such stimulants and bus loads of party-goers arrived at this night-club from as far away as Coventry and Rugby. Young adults were found wandering the streets in an apparently drunken state, often abusive and sometimes violent. Frequently individuals were arrested and, as the Police Surgeon, I would be called to examine the accused late at night. The offenders were all suffering from Amphetamine poisoning, which is somewhat similar to alcoholic intoxication. The Chief Inspector and I made detailed notes on every case and eventually we succeeded in helping the Authorities close down the night-club. A sad ending to a bold and well-meant enterprise.

Chapter 14: Locum Ship's Surgeon

One day, whilst discussing holiday ideas with my father-in-law, a retired GP, he reminded me that he had worked as a ship's surgeon for a short time when he was younger and he recommended that I do the same. I well remember my first introduction to cruising when I was a 14 year old schoolboy and my class went on a two weeks' educational cruise to Denmark and Norway. We sailed on the troop ship, "Neuralia", with about 400 other schoolboys, mostly from Scotland. I thoroughly enjoyed this cruise - it was a memorable holiday. It was tough, for I had never been away from home before, and the facilities were pretty basic. We slept on hammocks and I managed to sling mine near a port-hole each night which helped to keep my asthma at bay. I remember the latrines - about eight of them - all on the deck outside. There was no privacy as none had doors. Interestingly, we were one of the last passenger ships to go through the Kiel canal. On both the Danish and German embankments, we could see gun emplacements and we were strictly forbidden to take any photographs. The Second World War was to begin soon. On our way back across the North Sea to Leith, a gale force 9 blew up and most of the boys were sea-sick. I was all right and perhaps this fact removed my fears of cruising in the future - as a ship's surgeon.

It was in the early 1970s that I applied for a holiday job as Assistant Ship's Surgeon and was invited by the Shaw Savill Line to attend an interview in London. My wife was asked also and we eventually found our way to the Shaw Savill office building at the docks in Tilbury. I entered the interview room, and the fellow behind a large desk immediately answered the telephone, apologising for the interruption. He then told me that there was a lady on the line asking if she could still go on a cruise even though she was 29 weeks' pregnant. He asked my advice whilst she held on and I quickly replied "No way". This must have been the correct answer and it was also the end of my interview! We had a chat about the weather and about life in general - and I got the job!

I cruised for two to three weeks at a time, and twice a year, for about four years, each time on the "Northern Star". She was an old ship of 27,000 tonnes, a crew of 800 and carrying 1,200 passengers. My post as Assistant Ship's Surgeon was to be the Medical Officer for the crew. My Chief, the Senior Surgeon, looked after the passengers. He was paid a basic salary by Shaw Savill and charged a fee for all passenger consultations. I received no salary but had a free passage, and my wife and daughter received their cruise for half the brochure price. I had to sleep in a single cabin on the top deck in the officers' quarters, whereas the Senior Surgeon had a large, well furnished cabin with a double bed within the ship's hospital, and near to the nurses' quarters. The nurses' corridor was known as "Fluff Alley". The mind boggles! The two nurses on board had to be over 25 years of age and very well qualified, one being a trained midwife. Next to the doctor's cabin, was the consulting room, then an extremely well-equipped operating room, beyond which was a ward with four beds.

On the whole, I enjoyed my life on board ship. My surgery consultations were usually un-dramatic and routine, dealing with sea sickness,sore throats, coughs, earaches and tummy upsets. I had a few abscesses to open, using local anaesthesia, and one crew member, an electrician, was suffering with a very painful lower molar tooth. Extraction was the only treatment. I knew the procedure to give a block local anaesthetic, but first consulted my Hamilton Bailey's Emergency Surgery book, which was one of the text books I always carried in my case. The tooth, fortunately, came out intact. I was surprised at the amount of effort involved in using forceps and levers. Anyway, the electrician was so delighted with the outcome that he invited my wife, daughter and myself to a party held in the lowest deck of the ship for crew only and no officers. We all had a whale of a time!

A human problem presented itself at a morning surgery, one which I had never encountered before. A young girl who danced in the chorus came to see me. Upon sitting down she burst into tears. I muttered consolingly and she confided that her lesbian lover of five years had left her for another. This ex-partner was the lead singer - very tall, very attractive and with a lovely contralto voice. On board ship, such a situation can be completely devastating because there is no escape. I encouraged this unhappy girl to talk to me when she felt the need to unburden herself, gave her a few tranquilliser tablets as treatment, no more than two or three at a time. Unfortunately, a GP surgeon has no professional training to help with emotional problems, all I could do was advise like a Dutch Uncle.

Fractured bones were more in my line, and we had a few of those, especially in high seas. On one particular cruise, my Senior Surgeon was not versed in orthopaedics, so I had to deal with any broken bones. One fractured wrist of an elderly lady had quite a deformity - this was x-rayed by one of the nurses with a portable machine. I manipulated the fracture using a local anaesthetic, and with success. One other fracture involved the neck of the femur, again of an elderly lady, and this time we referred her to hospital in our next port of call.

On another cruise, we had two deaths. One was a woman who had been very jaundiced when she came aboard. Her own doctor had suggested a cruise, but unfortunately her cancer of the pancreas was very malignant. The other was an elderly man who had a sudden heart attack - apparently the third in that year. Both were buried at sea and the services, taken by the Captain at 0600 hours, were very moving. Only senior officers and relatives were present.

On my second cruise, I was presented with quite a challenge. I had just seen my wife and daughter into the embarkation lounge and was heading for the special gangway reserved for crew only when a call came over the tannoy, "Will the ship's doctor please report at once". I hurried to find that a child with a rash was just about to come aboard. I diagnosed chicken-pox, and at its most infectious stage. I had to refuse the family permission to embark. They were offered another cruise which they accepted with good grace, despite their obvious disappointment.

It was on my third cruise that I really walked into trouble. I felt alarmed when I first met the Senior Surgeon because I found him positively very inebriated. The following day, I was summoned to attend the Captain's cabin. This was unusual and I wondered what was up. I soon found out. All

135

the senior officers were sitting at a table and I was asked to join them. The captain said they were facing a crisis and hoped I could help. Furthermore, he told me he had just sacked the Senior Surgeon because of his persistent drunkenness, and formally asked me to take over the position. I agreed, and felt capable of dealing with the challenge. I was not to receive any salary and any fee charged to the passengers was to be part of the ship's income. From then on, I became a very busy man, even having to deal with the occasional night call. I did manage to get some off-duty time to visit the sights in ports and my wife and daughter were well looked after by the crew. Fortunately, on that particular cruise, I had no serious cases to deal with. During the voyage, my family and I gave a cocktail party in the Officers' Mess. This was very well attended and lots of canapés and a varied selection of drinks were laid on. We had four cheerful and efficient waitresses and we offered them drinks after the guests had left. It was then that I recognised one of them, which gave me a surprise, but I made no comment. The "waitress" was one of the male crew who sometimes served in the Officers' Mess at breakfast. I may add that he was very well made up and looked most convincingly female. Homosexuality was accepted on board ship, but not on dry land in those days.

One of my tasks was to inspect the ship for cleanliness and general hygiene. I was accompanied by the Captain and four Senior Officers. In the dining rooms, we lifted the odd piece of cutlery and, if any fault was found, the waiter in charge of that table was carpeted. Visiting the vast deep-freeze store was very uncomfortable, it was intolerably cold! We also inspected the crew quarters on the lower decks of the ship. In one cabin, a monkey was kept, another held a parrot and, strangest of all, one cabin was festooned with netting where finches flew around. This ship was home to many of the stewards. As we inspected each deck we ended up having a drink and, by the time the tour had finished, I was more than a little high, I can tell you! These officers certainly knew how to take their liquor.

On one of my voyages to the Western Mediterranean, my Senior Surgeon was a most delightful and amusing retired Rear-Admiral. He was a London Fellow and had been in the Royal Navy all his life. He seemed very taken with the two excellent nurses with whom he had sailed before. One morning, he asked me to see a girl of 13 years old complaining of severe abdominal pain. Upon examining her, I diagnosed, as he had done, acute appendicitis which, to our concern, was about to burst and cause peritonitis. We were two days from the nearest port and so decided that immediate operation was the correct course of action to take. We discussed the diagnosis and treatment with the girl's parents and their necessary consent was given for the operation to proceed. The Captain was quickly informed of the situation and he reduced the ship's speed down to two knots. Time was of the essence and we proceeded without delay. I gave the anaesthetic - fortunately with the old Boyle's machine which I knew well but had not used for years. It was a real challenge, for neither of us had removed an appendix for some time! My boss started the operation and exposed a very septic and inflamed appendix which was lying very deeply within the peritoneal cavity. He found that the severe deformity of his right hand (Duypytrene's contracture) prevented him from removing it. He asked me to take over. I handed the anaesthetic over to the senior nurse, scrubbed up, donned surgical gloves, and with some considerable difficulty secured and delivered the appendix without bursting it. We all experienced huge relief and the wound was sutured. The patient was prescribed a large dose of penicillin and made an excellent, uneventful recovery, much to the relief of the parents. The captain was informed that the operation had been a success and the ship's speed returned to normal. My boss gave me £5 and we all celebrated with a good dram of whisky.

Sea-sickness is, of course, a very common complaint, and much feared by some passengers. It is simple to treat, for in the majority of cases an injection of phenergan or stemetil at the first sign of sickness was the answer, coupled with anti-sickness tablets such as stemetil, stugeron or avomine. The latter I often gave for morning sickness suffered by ante-natal patients.

One year, when I did two cruises, the "Northern Star" docked as usual in Southampton for two days before her next departure. This time it was suddenly announced that the ship was to be visited by H.M. The Queen Mother, who had launched the ship 25 years before. Only the officers were to be on board, even my wife had to leave together with the Captain's wife. I found the Queen Mother such a delightfully charming lady. She had a firm hand grip when I was presented to her in the ship's hospital. She was so interested in everything and picked up some surgical instruments, wanting to know their names and uses in an operation. She had already heard about the emergency appendectomy on the previous cruise. The Senior Surgeon had joked beforehand about the possibility she might slip on the steep steps leading to our lower deck and injure her ankle. He wondered whether we might get a decoration for treating her! Luckily, no such thing happened. We all had lunch at the Captain's table. The loyal toast was given by the Chief Engineer, a Scot, who held his glass up high and said, "To the Queen Mum." She loved it! What a generous and popular Royal she was.

After completing four years as a Locum Assistant Ship's Surgeon, I called it a day.

Chapter 15: New Developments and Retirement

Our practice was allowed one district midwife, two district nurses and one Health Visitor. As time went by, I made increasing use of these practice colleagues. District nurses, however, were liable to leave for some reason or another, and had to be replaced. Latterly, I found an excellent S.E.N. (State Enrolled Nurse) who was fully committed and who had a great rapport with patients. I employed her increasingly to do follow-up calls for me and also to visit my geriatric patients. She enjoyed this extra responsibility, and she released me to deal with more urgent medical problems. I really did feel that the district nurse of the future could well be involved in more general practice work. We first appointed a Practice Nurse in 1965. Within five years, we had two and they were an enormous asset. I trained them in a few surgical techniques, such as taking blood samples and doing cervical smears. We also encouraged them to attend special courses for Practice Nurses. My partners encouraged me to do minor surgery, which of course I much enjoyed. This mainly involved injecting varicose veins, removing cysts and warts and dealing with ingrowing toe nails. There was no extra remuneration for these procedures, but, in later years, GPs with some surgical training were paid, and rightly so, because in tackling this minor surgery, the patient was treated swiftly and hospital waiting lists were reduced.

Over the years, the treatment of illnesses, injuries and disabilities changed dramatically for the better. The introduction of new drugs - beta blockers, antibiotics, tranquillisers and corticosteroids, to mention a few, increased and transformed the field of therapeutics. In early years in practice, one so often had to rely on the skill of one's hands in making a diagnosis. Taking a careful history was, and still is, of paramount importance. Biopsies became a more frequent method of investigation. Here, a tiny portion of the suspect lesion would be sent for histological (microscopic) investigation by a pathologist. Biochemical tests on body fluids also became more advanced and exact. Scanning investigations became major aids in making accurate diagnoses, and still later M.R.I. (Magnetic Resonance Imaging) was introduced. These latter investigations were in their infancy when I retired in 1984.

Another development in our practice was the introduction of a Family Planning Clinic. I went with a colleague for training at The Radcliffe Infirmary and we were both duly awarded the Certificate of Family Planning. We learnt the technique of fitting the diaphragm and the intra-uterine coil. These methods of contraception were, of course, later outdated by the introduction of the contraceptive pill.

The operation of vasectomy was then introduced in the early 1970s. I went to the first private vasectomy clinic in Swindon, run by a retired surgeon. He was a very helpful and skilled teacher and taught me his method. I decided to set up my own clinic within our surgery. Buying four sets of instruments was quite an outlay, but this number was required to deal with two or more clients per session. The instruments were sterilised for me at our local hospital and delivered ready to use. I was able to charge for each vasectomy, this helped our practice income and was a useful service for our patients. Other doctors in the area sent their clients to me and I enjoyed the extra surgery, as did the Practice Nurses.

In our town there was a small Nursing Home for N.H.S. and private patients, run very efficiently by Matron and Sister, both spinster ladies and both very kind to their elderly charges. Another GP and I looked after most of their patients. Ken was a close friend of mine and practised at another partnership in the town. Matron and Sister always had a Christmas party for their staff. Ken and I, together with our wives, were always invited. This celebration was held in a large hotel room and it was a hilarious event culminating with balloons, streamers and bangers being thrown around and Sister crawling on her hands and knees under the tables. We were a happy throng!

At one stage of my career, I had become very interested in the art of manipulation, or osteopathy. In those days, such treatment was frowned upon by the medical profession and we, as doctors, could not associate ourselves with an osteopath or chiropractor. Fortunately, this attitude was to change in later years.

I first became aware of the value of manipulation when I was surgical registrar to Professor Smilie at Larbert Orthopaedic Hospital in 1945. He quite often manipulated stiff joints and the spines of patients suffering with back pain. These procedures were usually carried out under anaesthetic. My partnership encouraged post-graduate study and the N.H.S. paid most of the expenses incurred. So, when a course in Manipulative Surgery came up in Poole, Dorset, I enrolled. It was brilliantly run by Professor Cyriax who was the first Professor of Orthopaedic Medicine at St. Thomas' Hospital, London. A number of GPs and some physiotherapists attended. At the end of the course, we had the option of sitting a professional examination, which the majority elected to do, including myself. Successful, I was given a certificate signed by the great man himself - Professor Cyriax. As well as instructing us in the art of manipulation, he also demonstrated how to inject painful joints. The injection was usually a local anaesthetic and a cortico-steroid preparation. He frequently performed these injections and manipulation techniques on actual afflicted patients. It was fascinating.

In my practice, I used my new manipulation skills on many occasions and often with dramatic results. My partners often referred cases to me. I recall the odd patient who suddenly found that he/she could not get out of bed or even move because of severe low back pain. I would first of all carry out a careful check on the nerve supply - if evidence of nerve injury was found, then manipulation was contra-indicated. When I could proceed, it was so very satisfying to relieve the patient of their painful disability or to watch them walk once more with a straight back. I often cured Tennis elbow by a simple manipulative procedure. "In manu vis medendi" - there is strength in thy healing hands.

I shall not forget one case of a lady with a whiplash injury to her neck, following a car crash. She made an appointment with me for manipulation, and one of my junior colleagues asked if he could observe the procedure as a matter of interest. The patient was duly made to lie on her back on the old leather treatment couch in my consulting room. Her head and neck supported in my hands were over-hanging one end of the couch. At the other end, her legs were held by two practice nurses who knew the drill. As usual, I

143

started to exert increasing neck traction, timed by one nurse. I twisted the patient's neck to the right and gave an extra jerk. Then I did the same to the left side and when I gave that extra jerk there was a sudden loud and dreadful crunching noise. Straightaway there was a crash behind me - my junior partner had fainted! The awful crunch had come from the leg of the old couch and my partner had thought that I had broken the woman's neck! When he revived, we all had a good laugh, including the patient whose injury was cured. I had great belief in the value of manipulation. Osteopathy and chiropractice, along with acupuncture, have now become accepted methods of alternative medicine.

Before I finish my story, I would like to write a little about my parents, who were both very dear to me. My father developed a heart block and was one of the first to be given a pace-maker. This was in the mid 1960s and thus made him a very important case study for cardiologists and medical students at Aberdeen Royal Infirmary, where he had received his life-saving operation. The pace-maker ensured that his heart rate was kept at 72 beats per minute - the normal adult figure. He led a reasonably active life on his farm for a further ten years. Then my mother, who later became disabled, moved to a Nursing Home near Banbury. She had always longed to visit the Outer Hebrides and often begged my father to take her there. But the nearest she got was Scourie, on the west coast, during our family camping holidays when I was a child.

When she died, we arranged for her body to be flown back to Scotland and to be buried with father. The hearse was waiting at Inverness airport, ready to take her on to Elgin.

However, by some oversight, the coffin was not unloaded at Inverness and she continued to fly on to the Isle of Lewis! The airline's mistake was discovered and she was returned to Inverness to be met by the patiently-waiting undertakers. And so, my mother attained her life-long wish. She did, after all, get to the Outer Hebrides, albeit a bit late!

My desire to retire came when I was just three months short of 65. I told my partners that I thought it was time the "old man" retired. They were very supportive and asked me to stay on a bit longer, but my mind was made up. It was a wrench and I may add, an emotional time, for I had to say "goodbye" to many patients whom I had looked after for so many years. I wondered how some of them would fare - especially those who were very elderly and infirm, a few hanging on to life by a thread. However, I had full confidence in my partners and in Gordon who followed me as Senior Partner. As time went by, I was pleased to learn that Tim, who was appointed to take over my list, turned out to be a worthy successor. Furthermore, I was very happy that the general management of our practice remained in the capable hands of Miss M. She had been a valued member of our team for nearly 40 years.

Retirement from an active professional life is both a physical and psychological experience. The sudden dawn of a new phase in life always comes as a shock. I was lucky to have made some plans for this change, as I had learnt from dealing with patients who had almost suffered a bereavement when faced with retirement. I often advised them to take up a hobby or sport, to join a Society, to give time to a worthy charity or to take part in local activities. As for myself, I loved country sports. I was already a member of the local golf club and I had enjoyed playing with many good friends. Now I was able to take up fishing, which I found very relaxing.

My favourite sport had always been shooting - after all, I was a farmer's son. I joined a shoot near Banbury and was to be a member of this syndicate for 24 years. Our bag of shot pheasants was not very good, but the company was wonderful and the countryside always a deep source of pleasure.

However, I was not yet destined to settle into full retirement, as after a few months, I was persuaded to return to work as a locum. This was not too onerous and did help to alleviate the sudden feeling of loss in not treating patients any more. I enjoyed this semi-retirement and welcomed the relief of having no more administrative responsibility to shoulder. Now, I mostly held surgery consultations with very few home visits. The introduction of computers in General Practice was fast approaching. No actual GP program was introduced until 1986-87. I learnt how to use the computer but I remained less than enthusiastic about this innovation.

Unfortunately, soon afterwards, my happily married life was indeed marred when my wife developed the awful, slowly progressive Alzheimer's Disease. She died about 8 years after presenting the signs of this malady. I found living by myself a challenge indeed.

The certain circumstances of loneliness and bereavement can thankfully be partially overcome by the heaven-sent gift of memory. One can so easily recall the images of one's lost companion and of the many happy moments of the past.

Socrates once said that no one appreciates life as much as an old man. Breathlessness and short windedness are common symptoms and synonymous with old age - "the braes (the hills) aye get steeper". But retirement presents for the elderly an opportunity for meditation and reflection. Perhaps one may find delight in the peace and quietness of an empty church or a secluded chapel. Or maybe relaxing in one's sitting room enjoying the tranquillity and listening to some gentle music. Or even possibly on board a cruise liner, gazing out on a calm sea whilst leaning over the side-rail.

Now, growing older, I am reminded of the expression, "You are as old as you feel, and if fit and well, then fifteen years younger than your actual age". Neither infirmity nor senility is synonymous with advancing years.

Going to church regularly has been without doubt a tremendous comfort and solace for me. I have made many new and now dear friends, socialising and keeping close to one's family is fundamental. To keep active physically and mentally is the prerequisite for a happy retirement.

I was fortunate to later re-marry. I met Ann on the cruise ship "Black Watch", sailing to the Canary Islands. We had both been widowed for 5 years. Neither of us was consciously seeking another partner, so we were both taken aback by the force of our feelings. Our children were delighted to witness our blooming romance. Ann and her dear, late husband had also been blessed with a son, Alistair and daughter, Beth.

We have found great happiness in sharing our twilight years together. After all, age does not protect you from love, but love, to some extent, does protect you from age.

It was after considerable thought and deliberation that I decided to move away from Bloxham and live in Ann's house in South Wales. It was a wrench leaving my many friends and all my associations with old patients. Caerleon, near Newport, is an attractive and particularly interesting town with strong archaeological links with its Roman past. Some claim it has connections with King Arthur and The Round Table. Here I have made good new friends and life has taken on a new dimension. And we so enjoy walking our Staffordshire dog, Cora, in the beautiful Monmouthshire countryside.

I shall conclude with a paragraph by John Aubrey which I frequently read and which I copied on to the first page of all my annual GP visiting diaries:-

"Do not agitate yourself over anything, do all things tranquilly and in a spirit of repose, for nothing whatsoever lose your interior peace, even if everything be turned upside down; for what are all the things of this life compared with peace of heart".

~ ACKNOWLEDGEMENTS ~

Many of my friends, patients, and especially my family, have encouraged me to write this account of my working life.

I am most grateful to my daughter, Fiona, who has helped so much in the word processing and checking of my script, let alone managing to decipher my writing - yes, it **is** true what they say about doctors' writing!

I am indebted to my wife, Ann, with whom I spent many hours re-reading and editing the book, which resulted in a mixture of laughter, sadness and reflection.

My thanks also to Alistair, whose help has been invaluable and to my publisher Eamonn who has been particularly supportive and encouraging.

However, the real stars are, without doubt, the thousands of patients I had the privilege of treating during my career, as without them this book would not have been possible. I thank them for putting their trust in me.